CONTROLLING FRUIT FLIES BY THE STERILE-INSECT TECHNIQUE

PROCEEDINGS OF A PANEL
AND RESEARCH CO-ORDINATION MEETING
ORGANIZED BY THE JOINT FAO/IAEA DIVISION OF ATOMIC ENERGY
IN FOOD AND AGRICULTURE

INTERNATIONAL ATOMIC ENERGY AGENCY, VIENNA, 1975

CONTROLLING FRUIT FLIES
BY THE STERILE-INSECT TECHNIQUE

PANEL PROCEEDINGS SERIES

CONTROLLING FRUIT FLIES
BY THE STERILE-INSECT TECHNIQUE

PROCEEDINGS OF A PANEL
AND RESEARCH CO-ORDINATION MEETING
ON THE STERILE-MALE TECHNIQUE
FOR CONTROL OF FRUIT FLIES
ORGANIZED BY THE
JOINT FAO/IAEA DIVISION OF ATOMIC ENERGY
IN FOOD AND AGRICULTURE
AND HELD IN VIENNA, 12-16 NOVEMBER 1973

INTERNATIONAL ATOMIC ENERGY AGENCY
VIENNA, 1975

CONTROLLING FRUIT FLIES BY THE STERILE-INSECT TECHNIQUE
IAEA, VIENNA, 1975
STI/PUB/392
ISBN 92−0−111575−X

Printed by the IAEA in Austria
November 1975

FOREWORD

The effectiveness of the sterile-insect technique (SIT) was amply demonstrated even before the recent steep rises in oil prices. With the increasing cost of oil-derived insecticides, the SIT becomes all the more important since the efficient use of this technique depends on minimizing the application of insecticides and maximizing the utilization of local resources. In this context the SIT is an effective and environment-sparing alternative to indiscriminate use of chemical control, and is especially significant for developing countries with their limited foreign exchange and still relatively unpolluted environments. Many of these countries rely heavily on agricultural exports of fruit and fruit products for their foreign exchange. Appropriately, most SIT campaigns up to now have been directed against fruit flies.

Since the previous FAO/IAEA panel on the control of fruit flies by the SIT, published by the IAEA in 1970 under the title "Sterile-male Technique for Control of Fruit Flies", much progress has been made and aspects of the SIT so far ignored or neglected have come to receive increasing attention. Among these aspects is quality control by flight-mill studies, mating speed, and measurement of genetic distance between populations by allozyme analysis. In addition, new field operations using sterile releases and related techniques have been carried out against such pests as medfly, melon fly, cherry fruit fly, olive fly and onion fly. A panel and research co-ordination meeting to survey these developments was held by the Joint FAO/IAEA Division of Atomic Energy in Food and Agriculture in Vienna on 12–16 November 1973 and the present book contains the proceedings of that combined meeting.

Since then insecticides have become even more costly, and the containment of fruit flies by methods less reliant on insecticides has taken on a new urgency. The need for such fruit-fly containment and for its timely application can be strikingly illustrated by an example taken from Central America. The Mediterranean fruit fly, which first appeared in the 1950s in Central America, has surged northward throughout this area and is now poised to invade Mexico. The economic significance of an unchecked spread of this pest is reflected in an estimated annual loss of $6.8 million for the Central American area excluding British Honduras and Mexico. Should the medfly invade those two countries, their citrus industries would suffer a combined annual loss of $6 million. Should the medfly go unchecked and invade the United States, the US fruit industry would lose annually about $85 million for citrus and $200 million for deciduous fruits. These figures come from "Economic Survey of the Mediterranean Fruit Fly in Central America", USDA/USAID Field Report No. 21, by R.G. Henning et al. (1972).

This situation in Central America has revived interest in the Joint UNDP/OIRSA/IAEA co-operative project. Seven countries in this area collaborated in the project from 1965 to 1970; although the outcome was a successful demonstration of medfly suppression by the SIT, there was unfortunately no practical sequel to it. In 1974, remedying this situation would have required the investment of an estimated $20 million in a five-year campaign of eradication culminating in area-wide releases of sterile medflies. In spite of this heavy cost, eradication was still at that time considered a feasible and economically possible proposition. Now, however, in 1975, prices have rocketed so steeply and the pest has spread so widely that eradication may no longer be economic or practicable, and containment including the use of guardian populations of sterile flies may be a last resort. Because of its particular relevance this book includes as an appendix a paper by R.H. Rhode, originally planned for the 1974 IAEA Symposium in Innsbruck on the Sterility Principle for Insect Control. This paper describes some of the organizational, strategic and technical aspects of what, under Central American conditions, has become a formidable undertaking.

It is hoped that the present book will be a useful addition to the series of panel and symposium proceedings published by the IAEA on the sterile-insect technique, and will encourage redoubled efforts in the use of this non-polluting method of controlling fruit flies.

CONTENTS

STATEMENTS AND RECOMMENDATIONS

PRINCIPAL PAPERS

THE STERILE-INSECT TECHNIQUE
FOR THE CONTROL OF FRUIT FLIES

A survey

E.J. HARRIS
Hawaiian Fruit Flies Lab.,
USDA, ARS,
Honolulu, Hawaii,
United States of America

Abstract

THE STERILE-INSECT TECHNIQUE FOR THE CONTROL OF FRUIT FLIES: A SURVEY.
Some advantages of the sterile-insect technique (SIT) are its minimum contribution to environmental pollution and its minimum adverse effect on non-target organisms. A review is made of the melon fly and sterile Mediterranean fruit fly release programmes, the accomplishments, and the implications. Recommendations are made for research leading to development of methods for practical use of the SIT.

INTRODUCTION

In the United States in recent years, the sterile-insect technique (SIT) has emerged as an important approach to biological control of pest insect species and an alternative to chemical control. This technique (here, this term means the release of sterile male and female insects) of controlling fruit flies (Tephritidae) was adopted early in Hawaii (Steiner et al., 1962, 1965), not long after Knipling reported on this bio-control principle and Baumhover et al. (1955) demonstrated its practicality. Subsequently, other tests of the technique were made by Cheikh et al. (1975), de Murtas et al. (1970), LaBrecque and Keller (1965), Lopez-D. (1970), Mellado et al. (1970), Monro (1965), Rhode et al. (1971), Shaw and Riviello (1965), Steiner (1969), and Steiner et al. (1962, 1965).

The duty of this panel is to review the research involving SIT, to note the progress made, and to determine what needs to be done to accelerate progress toward the ultimate goal, the routine use of the SIT for practical control of fruit flies.

PROGRAMME REVIEW

Trends toward integrated control (Stern et al., 1959) and pest management programmes that involve the use of parasites, predators, and insecticides have recently become pronounced (Huffaker, 1971). This approach is a logical development since it should avoid or minimize the development of insect resistance to pesticides, reduce the amounts of pesticide used, and avoid the elevation of secondary pests to the status of major pests as a result of the destruction of beneficial insects. However, effective integrated control and pest management require basic information about the ecology and seasonal populations of the target species, about their parasites and predators, and about other elements in the insect fauna. Likewise, the eco-system of the pest insect must be monitored constantly during control operations to determine or predict effects on the pest species and other species. This same information and thinking is also mandatory when the SIT is to be used. Indeed, almost all intelligent methods of insect control require ecological knowledge, but the SIT normally requires very much more, for example, data on dispersal, migration, mating habits, and density.

There are few disadvantages to the SIT if it is used when it will work. For example, it was used successfully by the Hawaiian Fruit Flies Laboratory to eradicate the melon fly, Dacus cucurbitae Coquillett, from Rota in 1963 (Steiner et al., 1965). Also, eight incipient outbreaks (undoubtedly started by cyclonic winds carrying flies from Guam 37 miles away) have been stopped on Rota during the last 10 years. These successes are good evidence of the effectiveness of the technique when it is used properly: the method will eradicate pests and will contain outbreaks at minimum cost. However, the technique has not always been successful against the melon fly. For example, when it was used in Guam between 1969 and 1972 (Government of Guam, Department of Agriculture, unpublished data), the sterile flies proved to be non-competitive, and the monitoring system inadequate; also, the topography of the island was such (ecological niches in which large numbers of the fly could breed undetected) that reinfestation of cleared areas occurred quickly.

The success of the SIT against the Mediterranean fruit fly, Ceratitis capitata (Wiedemann), also depends on careful planning. However, interest in suppressing this species is high because of the medfly's worldwide economic importance. As a result, releases of sterile medflies have been made in Hawaii, Italy, Spain, Cyprus, Peru, Argentina, Australia, Chile and Israel, and, more recently, in Costa Rica and Tunisia. The most important of these tests, because of the size and the potential for practical control, were those made in Costa Rica and Tunisia. The efforts of Rhode et al. (1971) achieved successful suppression of the medfly in Costa Rica. Also, Cheikh et al. (1975) reported satisfactory results against the medfly in Tunisia.

The SIT is young and is evolving. This evolution must proceed in a reasonable way. Let us therefore look at these most recent efforts to use the technique to see what lesson we can learn.

EVALUATION OF RECENT PROGRAMMES

In both Costa Rica and Tunisia the original experimental area was quite large, but in each case, after 1 or 2 years, it was reduced to a size that could be adequately overflooded with the available sterile medflies. After the experiment had been carried out the original estimates on the resources needed to handle the experimental areas were seen to be unrealistic. Unfortu-nately, methods do not exist at present for obtaining in advance completely realistic estimates of what is needed to implement such large-scale pro-grammes. Even a success with the technique in a small area does not necessarily provide the information required to set up a large-scale test. Such difficulties cause some critics to say that the SIT is not practical. However, I do believe that the method is practical and that it can be used efficiently when rearing and release techniques are correct and ecological conditions are right. In the case of sterile-insect release projects in Costa Rica and Tunisia, the rearing programmes were effective after some minor problems were worked out; much credit for this is due the laboratory personnel who were enthusiastic and careful; also, it is comparatively easy to check that reared insects are developing well. The question of the right release techniques and the resulting data (trap recoveries and fruit sampling) is not as easy to check. Similarly, ecological conditions that depend on the co-operation of large numbers of people (like sanitary measures) are difficult to predict or achieve in any such large-scale tests.

The successes in Costa Rica and Tunisia were limited in both duration and scope. However, the experiences greatly aided the development of the SIT. We must obtain such base-line information before we can proceed to the next level of the development process; we must profit from past mistakes; and we must use all the data obtained in planning subsequent large-scale tests.

RECOMMENDATIONS AND FUTURE RESEARCH

Obviously, large-scale tests of the SIT are expensive, and favourable results cannot be guaranteed. However, chemical treatments for pest control fail too sometimes — perhaps because of rainfall, incomplete coverage, drift, or insect resistance. To correct for difficulties of this type, we try to co-ordinate treatments with the optimum weather conditions, give better coverage of the crop, or reduce retreatment intervals. Failures with pesticides are therefore accepted and not usually attributed to the method — like they are with the SIT. Likewise, failures with released parasites and predators are accepted on the basis that a good system of evaluating and maximizing their effectiveness has not yet been developed; the SIT deserves similar forbearance.

To develop the system necessary to succeed with the SIT, we should co-ordinate our efforts, we should share information, and we should minimize duplication of effort. Also, we must recognize that in small-scale tests with sterile medflies, we can control the programme and study all necessary parameters. In the large-scale tests, we cannot control every detail, and we can monitor only the most important parameters. Also, it seems that when the size of the release area doubles, the resources needed apparently quadruple. A major area of future research will therefore involve the determination of the really essential parameters for a given test. We know that they vary with the location and test conditions. In Hawaii, the choice of test site is not simple, and pretreatment to suppress the native fly population is necessary. Also, the average population level is high and the climate is favourable for the medfly so success is difficult in Hawaii. However, even if we do not have the most favourable climatic conditions in Hawaii and do have many favourable inaccessible areas in which the fruit flies breed, we still have areas where we can conduct large-scale tests with sterile releases alone or in combination with poison baits and male-annihilation treatments.

Nevertheless, we have not made sufficient studies of the ecological relationship between fruit flies and other insect fauna in the environment. For example, as noted, the ecosystem in Hawaii is characterized by a climate that is favourable for fruit flies but not particularly favourable for dependable control operations, and topographic conditions that make access difficult

to many areas. However, on Tinian in the Mariana Islands, where the eco-
system is simpler than on Hawaii, the melon fly disappeared in 1963 despite
apparently favourable conditions; we do not know why. Since we should
implement sterile fruit fly release programmes when natural factors are
exerting their maximum suppressive effects on the native fly population,
we must know when these factors are exerting their maximum effects.

In Israel, Tunisia, Morocco, and probably Algeria, the climate in the
citrus-growing areas is such that the crop must be protected in summer
and fall to reduce damage to the fruit from females and from larvae; and
such protection is by chemical means. As a result, in contrast, in Tunisia,
the medfly problem is particularly severe in those locations where fruits
that mature at various seasons of the year are grown near citrus; also the
land area used in citrus production is smaller, and there is more intensive
cultivation of this land, which aggravates the problem. (When citrus is
grown in large pure and isolated stands, the medfly problem is less severe.)
Bait sprays, the immediately effective method, and the SIT, the long-term,
long-lasting, effective method, may therefore be compatible in Tunisia and
these other countries but priority must be given to meeting immediate needs.

In addition, the SIT may not work against all fruit fly species or even
against the same species in all the areas of the world in which they are
found. This is especially true if a system of reducing populations to a
manageable level and if accessibility to all favourable ecological niches are
not possible.

Finally, in my opinion, priority should be given to the development of
pheromones, female lures, and a more efficient method of monitoring population trends and male annihilation of the medfly. Such attractants are
specific and could be applied with conventional spray or special equipment
in combination with the SIT; they would also have a minimum polluting effect
on the ecosystem and would permit the use of sterile flies to mop up pockets
of survivors left by the attractants. The fruit fly parasites could be mass
reared and released separately or simultaneously with sterile fruit fly
releases. This approach may function better in Hawaii and Central America
than in Europe.

In closing, I list for consideration the following work as worthy of
further effort in a co-operative programme.

1. Develop methods of determining the essential ecological parameters
that affect fruit fly populations. Remote sensing may be used to determine plant
host identity, density, and distribution, and computer population models may
be used to predict serious outbreaks.

2. Develop techniques to integrate chemical and cultural control methods
with the SIT. Non-preferred hosts for human consumption may be used as
trap crops in a cultural control to collect and destroy fruit flies during
unfavourable periods.

3. Conduct co-ordinated research to avoid duplication of effort and
accelerate development of (a) quality control techniques, (b) sterile release
methods, (c) strain development, (d) assessment of sterile release effects,
and (e) assessment of sterile fly behaviour in comparison with wild flies.

4. Determine when and how to apply integrated control (pheromones,
female lures, and male-annihilation treatments) in combination with the
sterile-insect technique.

5. Develop methods of releasing fruit flies sterilized with chemosterilants.

6. Develop methods of mass rearing and releasing fruit fly parasites.

REFERENCES

BAUMHOVER, A.H., GRAHAM, A.J., BITTER, B.A., HOPKINS, D.E., NEW, W.D., DUDLEY, F.H., BUSHLAND, R.C. (1955), Screw-worm control through releases of sterilized flies, J. Econ. Entomol. 48, p. 462.

CHEIKH, M., HOWELL, J.F., BENSALAH, H., HARRIS, E.J., SORIA, F. (1975), Suppression of the Mediterranean fruit fly in Tunisia with released sterile insects, J. Econ. Entomol. 6, p. 237.

DE MURTAS, I.D., CIRIO, U., GUERRIERI, G., ENKERLIN-S., D. (1970), "An experiment to control Mediterranean fruit fly on the island of Procida by the sterile-insect technique", Sterile-Male Technique for Control of Fruit Flies (Proc. Panel Vienna, 1969), IAEA, Vienna, p. 59.

HUFFAKER, C.B. (1971), Biological Control, Plenum Press, New York and London.

KNIPLING, E. (1955), Possibilities of insect control or eradication through use of sexually sterile males, J. Econ. Entomol. 48, p. 459.

LaBRECQUE, G.C., KELLER, J.C. (Eds), (1965), Advances in Insect Population Control by the Sterile-Male Technique, Tech. Reports Series No. 44, IAEA, Vienna.

LOPEZ-D., F. (1970), "Sterile-male technique for eradication of the Mexican and Caribbean fruit flies: review of current status", Sterile-Male Technique for Control of Fruit Flies (Proc. Panel Vienna, 1969), IAEA, Vienna, p. 111.

MELLADO, L., NADEL, D.J., ARROYO, M., JIMENEZ, A. (1970), "Mediterranean fruit fly suppression experiment on the Spanish Mainland in 1969", Sterile-Male Technique for Control of Fruit Flies (Proc. Panel Vienna, 1969), IAEA, Vienna, p. 91.

MONRO, J. (1965), "Experimental control of Dacus tryoni", Advances in Insect Population Control by the Sterile-Male Technique (LaBRECQUE, G.C., KELLER, J.C., Eds), Tech. Reports Series No. 44, IAEA, Vienna, p.22.

RHODE, R.H., SIMON, J., PERDOMO, A., GUTIERREZ, J., DOWLING, C.F., Jr., LINDQUIST, D.A. (1971), Application of the sterile-insect release technique in Mediterranean fruit fly suppression, J. Econ. Entomol. 64, p. 708.

SHAW, J.G., SANCHEZ RIVIELLO, M. (1965), Effectiveness of tepa-sterilized Mexican fruit flies released in a mango grove, J. Econ. Entomol. 58, p. 26.

STEINER, L.F. (1969), Control and eradication of fruit flies in citrus, Proc. First Int. Citrus Symp. 2, p.881.

STEINER, L.F., ROHWER, G.G., AYERS, E.L., CHRISTENSON, L.D. (1961), The role of attractants in the recent Mediterranean fruit fly eradication program in Florida, J. Econ. Entomol. 54, p. 30.

STEINER, L.F., MITCHELL, W.C., BAUMHOVER, A.H. (1962), Progress of fruit fly control by irradiation sterilization in Hawaii and the Mariana Islands, Int. J. Appl. Radiat. Isotopes 13, p. 427.

STEINER, L.F., HARRIS, E.J., MITCHELL, W.C., FUJIMOTO, M.S., CHRISTENSON, L.D. (1965), Melon fly eradication by overflooding with sterile flies, J. Econ. Entomol. 58, p. 519.

STERN, V.M., SMITH, R.F., VAN DEN BOSCH, R., HAGEN, K.S. (1959), The integration of chemical and biological control of the spotted alfalfa aphid. The integrated control concept, Hilgardia 29, p. 81.

GENETIC VARIATION IN NATURAL INSECT POPULATIONS AND ITS BEARING ON MASS-REARING PROGRAMMES

G. L. BUSH
University of Texas,
Austin, Texas,
United States of America

Abstract

GENETIC VARIATION IN NATURAL INSECT POPULATIONS AND ITS BEARING ON MASS-REARING PROGRAMMES.
A new approach using gel electrophoresis of genetically variable enzymatic and non-enzymatic proteins (allozymes) which can be used to monitor laboratory populations for genetic changes is discussed. Examples of genetic alterations in laboratory strains of the codling moth and screw-worm fly are presented. The results of a survey of genetic variation in the European cherry fruit fly are compared with the codling moth. The technique is very sensitive to environmental change, and offers a simple, rapid method of quality control.

The success of any sterile male release program is obviously predicated on the ability of factory produced insects to locate and mate with their wild counterparts. Because mating behavior in insects consists primarily of fixed action patterns which are under tight genetic control, artificial selection under laboratory conditions can and frequently does produce striking behavioral changes in a few generations [6]. Regrettably, this important fact is too often given a great deal of lip service but ignored or relegated to a back burner in the development of mass rearing and production line techniques.

Quantity rather than quality has generally been the measure of success in most mass rearing programs. Quality control programs that actually test for the performance of the insect under field conditions are almost non-existent. This approach has led to some rather disastrous consequences as witnessed in the recent collapse of the screw-worm (Cochliomyia hominivorax) suppression and eradication program along the U.S.A. - Mexico border [5].

There are at least three reasons why effective quality control programs have not been incorporated into mass rearing programs like that of the screw-worm fly. First, until recently few individuals associated with mass rearing programs realized or appreciated the evolutionary implications and potential genetic pitfalls of adapting wild insects to the artificial conditions associated with factory production. Second, little is actually known about the critical features of reproductive behavior and ecology of our insect pests which is essential for the development of meaningful quality control techniques. Finally, we have only meager information about the genetic structure of natural insect pest populations and the behavioral and ecological genetics of specific traits involved in mating behavior in the field.

If we accept the premise that one important goal of any mass rearing program is to maintain the correct combination of allelic forms at loci essential for normal mating behavior, we are faced with the problem of how this can be accomplished when basic behavioral and ecological data as well as vital genetic information on the species in question is usually non-existent.

The screw-worm offers a good example of the problems facing most mass rearing programs developing effective quality control programs. Essentially nothing is known about the mating behavior of this diurnal fly in nature. Because no one has ever seen this species mating in the field and essential information on its ecology is lacking, no technique is available to monitor the production line flies for genetic changes affecting behavior.

The fact that the males have larger eyes than females and emit a pheromone [7] indicates that both visual and chemical cues are important in mating behavior. Yet in factory reared flies used for release last year (APHIS Regular Production Strain) at least three recessive eye color mutants occurred at a frequency of about .01 percent. In Drosophila these homozygous mutant forms are partially or completely blind [8] and, more importantly, individuals heterozygous for color mutations are generally less fit than wild flies [6]. In the screw-worm production flies approximately 2% of the flies are heterozygous for these visible subvital or sublethal alleles. More subtle deleterious mutations are also likely to be present in the lab population, but without adequate methods for estimating fitness their effects can not be established.

In addition, certain laboratory strains are known to differ in the female's response to male pheromone. Females of the Mexican Strain do not respond to Florida Strain male pheromone [8]. The response of wild females to laboratory strain male pheromone has not been tested, but if flies with aberrant pheromones are produced, the effectiveness of the production line flies would be further reduced.

Difficulty has also been encountered in getting wild screw-worm flies to oviposit and mate under laboratory conditions [11]. The solution to this problem in the past has usually been to use the few flies that mate under artificial conditions as progenators of the next generation. After five or six generations "normal" fertility and fecundity is restored. However, these "normal" flies are the very ones that are most likely to be genetically aberrant and may be useless for release purposes.

A look at the methods used in mass rearing screw-worms suggests why genetically aberrant individuals exist in the production line flies. Until recently adults were held in total darkness to permit crowding and thus reduce rearing costs. Flies usually walked rather than flew to their food source and the oviposition device. Visual and odor cues were relatively unimportant for survival in this environment and the selective advantage was conferred on those flies with genotypes that permitted them to mate and oviposit in the dark under crowded conditions. A blind fly is not at a disadvantage in such a colony as has been demonstrated in experiments with blind Drosophila reared in the dark with normal flies [9].

It is not surprising, therefore, that the mating success of the mass reared flies has deteriorated. Because artificial conditions are common to most insect mass rearing facilities, including those used in rearing such species as Dacus oleae and Ceratitis capitata, similar reductions in competitiveness are inevitable unless guarded against.

Some means must therefore be developed to ensure that genetic changes do not occur during laboratory adaptation. Before suitable bioassay techniques can be introduced we must know how much genetic variation occurs naturally at many different critical loci, and how fast and what kind of genetic alterations occur during domestication. Furthermore, we must know how important these changes may be to the competitive success of laboratory reared insects in the field.

Flight mill studies, mating speed, preference and propensity tests, startle reactions and several other methods (see Chambers, this Panel) have been proposed and studied as possible means of monitoring laboratory strains for behavioral changes. Because all of these measure some component of behavior under completely abnormal conditions, it is difficult to establish if the results reflect normal behavior encountered in field conditions even when using wild flies for the test. Furthermore, it is almost impossible to obtain estimates of the total amount of genetic variability for any of these traits. Usually some form of a value judgment must be made as to what constitutes normal behavior.

There is one way to circumvent this problem through the use of gel electrophoresis of polymorphic enzymatic and non-enzymatic proteins (allozymes). This technique can be used to monitor the level of genetic variation at several loci in both natural and laboratory populations. It was first used with insects by Lewontin and Hubby [12] and is now widely applied to the study of population genetics in several insect groups. A review of the technique as applied to several species of Tephritidae is available [4] along with a step-by-step outline of the methods employed [3].

Past work has dealt mostly with the genetic structure of wild populations, but recently I have adapted it for use in a mass-rearing quality control program for the screw-worm, codling moth, and a species of Tephritidae. Allozyme data can routinely be used to monitor laboratory strains for any indication of genetic shifts that might be going on at other loci which cannot be studied directly. We will look at the codling moth first.

In the codling moth (Laspeyresia pomonella) a genetic survey of 27 loci in natural and laboratory populations has revealed a pattern of variation typical of what one would expect in laboratory populations undergoing intense inbreeding and artificial selection (Bush,in preparation).

Figure 1 is a dendrogram of genetic distances that exist between various populations of codling moths. It was derived by an unweighted pair-group method of clustering developed by Sokal and Sneath [14], using generalized genetic distances calculated on the basis of weighted protein mobility rates of five polymorphic loci (PGM, EST-A, EST-C, ADH, AK-A). A copy of the program and summary of methods employed to calculate these distances, which were developed by Drs. P. Smouse and R. Richardson (Department of Zoology, University of Texas),can be obtained on request.

It is quite evident that all natural populations of the codling moth (those underlined in Fig. 1) share similar alleles and gene frequencies. Even the walnut and apple races, which are considered by some biologists to represent rather distinct non-interbreeding populations in California, show no statistically significant differences in gene frequency. Little or no correlation, however, is evident between the genetic structure of the laboratory reared populations and the localities from which they originated. This indicates that they have diverged genetically from the natural populations.

An analysis of genetic variability in natural vs. laboratory populations is also interesting (Tables I and II). At six polymorphic loci (PGM, PGI, EST-C, EST-A, ADH, and AK-A) the wild populations in Western North America on the average have 2.7 alleles per locus while the laboratory populations from the same regions (represented by an *) average only 2.43. This represents a loss of overall genetic heterozygosity and indicates that a certain amount of inbreeding or selection or a combination of both has occurred in the laboratory populations. This observation is further supported by the fact that of the loci sampled in laboratory populations, 12.5% were significantly deficient in heterozygotes indicating strong inbreeding. Of the loci sampled in the wild populations, only 2 (6.7%) lacked a sufficient number of heterozygotes.

A preliminary study of screw-worm flies gives very similar results although the data is not yet extensive. After a survey of 36 loci, the five most polymorphic loci were selected for analysis. Table III summarizes the

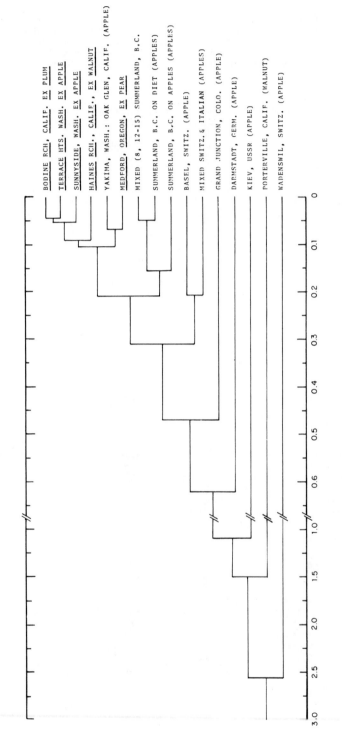

FIG. 1.   Genetic distance between natural (underlined) and laboratory populations of the codling moth.

TABLE I. NUMBER OF ALLELES AT SIX LOCI POLYMORPHIC IN
NATURAL POPULATIONS OF THE CODLING MOTH
Samples significantly deficient in heterozygotes with circles; those showing
a surplus of heterozygotes with squares

Wild populations	Number of alleles per locus						Average
	PGM	EST-A	EST-C	ADH	PGI	AK-A	
1. Sunnyside, Wash. (ex. apple)	3	5	3	1	2	3	2.83
2. Terrace Hts., Wash. (ex. apple)	3	5	2	1	☐2	2	2.50
3. Bodine Rch., Calif. (ex. plum)	4	⑤	3	1	2	2	2.83
4. Haines Rch., Calif. (ex. walnut)	3	3	3	1	2	2	2.33
5. Medford, Oregon (ex. pear)	⑤	5	3	1	2	2	3.00
Average	3.60	4.60	2.80	1.00	2.00	2.20	2.70

Loci with significant surplus of homozygotes = 6.7

findings. The number of alleles is not given as the genetics of each locus
has not yet been established. There is evidence, however, that even in laboratory populations polymorphic for the same locus the alleles involved may
be different. The level of differentiation between strains is therefore
greater than indicated.

The APHIS Regular Production Strain and the Puerto Rican Strain are
segregating only at the EST locus. The Regular Florida and NST Texas Strains
are both segregating at the EST and α-GDH locus while the remaining
four laboratory strains are segregating at other loci as well. We as yet
have only a few samples of wild flies reared from individual egg masses collected on wounded animals. However, there is a strong indication that the
wild flies are polymorphic at more loci and have more alleles per locus than
do their laboratory reared counterparts. Because of the screw-worm's ability
to disperse long distances, [10], genetically homogeneous populations probably
extend over broad regions with little genetic variation existing between
widely separated localities.

Evidence of genetic changes in laboratory colonies is not yet available
for any species of Tephritidae, but we do have extensive evidence on wild populations. The same methods were used to calculate the genetic distances as
were used on the codling moth.

Recent investigation of genetic variation in four polymorphic loci in
natural populations of Rhagoletis cerasi (Bush and Boller, in preparation)
has revealed that gene frequencies for this species do not vary substantially
over its entire range. In R. cerasi no significant differences could be

TABLE II. NUMBER OF ALLELES AT SIX LOCI POLYMORPHIC IN
LABORATORY POPULATIONS OF THE CODLING MOTH
Samples significantly deficient in heterozygotes with circles. Except
where noted all populations were reared on artificial laboratory diets

Lab Populations	Number of alleles per locus						Average
	PGM	EST-A	EST-C	ADH	PGI	AK-A	
1. Yakima, Wash.*	3	5	2	1	2	3	2.67
2. Porterville, Cal.*	3	3	3	1	2	②	2.33
3. Oak Glen, Cal.*	3	③	3	1	2	2	2.33
4. Summerland, B. C.* (Apple)	3	④	3	1	2	②	2.50
5. Summerland, B. C.*	3	④	3	1	2	②	2.50
6. Grand Junction, Colo.	5	④	2	2	2	2	2.83
7. Wadenswil, Switz.	4	3	2	1	2	2	2.33
8. Darmstadt, Germ.	④	②	3	1	2	3	2.50
9. Mixed Switz. & Italian	3	2	2	1	2	3	2.16
10. Basel, Switz.	2	3	2	1	2	3	2.16
11. Kiev, USSR	3	3	2	1	2	3	2.33
12. Mixed (4, 8-12)	4	3	2	1	2	3	2.50
Average	3.30	3.25	2.42	1.08	2.00	2.50	2.43

Loci with significant surplus of homozygotes = 12.5%

*Samples compared in text with natural populations in Table I.

found even between the eastern and western populations of the cherry infesting
race which show unidirectional sterility and probably represent distinct species
or at least semi-species [2]. Furthermore, even when individuals of the most
distant populations of R. cerasi (Spain, Italy, Turkey, the Netherlands, and
Germany) are included in the analysis (Fig. 2), the genetic distances are about
the same order of magnitude as exist between the wild codling moth populations
in western North America.

These close similarities in the genetic structure over broad regions
between reproductively isolated populations suggest that the frequency of
alleles at each locus is being tightly held in an adaptive balance by natural
selection. This lack of geographic variation at the allozyme level in insect
populations has been encountered by several investigators [13] and appears to
be related to the type of environmental heterogeneity (graininess) encountered
by most insects.

TABLE III. FIVE LOCI POLYMORPHIC IN AT LEAST SOME LABORATORY
POPULATIONS OF THE SCREW WORM FLY
Over 50 individuals were electrophoresed from each population reared at
the USDA Screwworm Research Laboratory, Mission, Texas
+ = two or more alleles present

Adults	Loci				
	LAP	ALD	PGM	GDH	EST
APHIS-Regular Production					+
PRN-F9-Puerto Rico Strain					+
NTS-New Texas Strain				+	+
Regular Florida Strain				+	+
C3-F5-Texas Collection			+	+	+
ADHIS-Old Mexico Strain		+		+	+
C9-F1-New Mexican Collection	+			+	+
TX-F8-New Texas Strain		+	+		+

It could be predicted therefore that any shift that might occur in laboratory populations of R. cerasi and other Tephritidae away from the normal
genetic structure of wild populations would provide an excellent indication
that selection is occurring in these laboratory colonies. Also, change at
any of these loci during the process of colonization would indicate that
shifts in gene frequencies might also be occurring at other key loci as well.
Steps could then be taken to alter rearing methods in order to halt any further
change and possibly to restore the population to its original state.

We are now developing a simplified sampling method and the computational
tools to apply the allozyme polymorphism approach to a mass rearing monitoring
system. This technique will not only provide an early warning system of genetic change, but will also furnish data on the rates at which these changes
occur and give some indication of the levels of inbreeding that are going
on in the laboratory population.

Because changes in allozyme frequency appear to be extremely sensitive to
environmental change brought about by mass rearing techniques, they can offer
a simple, rapid method of quality control. As with any existing method now
in use, allozymes do not measure directly those genes that are affecting some
critical phase of behavior. However, because they represent a direct measure
of the genetic structure of the population they are probably a much better in-
dicator of how effective mass rearing programs are in maintaining the natural
genetic structure of the wild populations than some of the other techniques now
in use. The allozyme method, particularly when used with other methods of quality control, can reduce the chance of selecting for unwanted genotypes that reduce the effectiveness of laboratory reared flies and provide an early warning
system that trouble may be ahead.

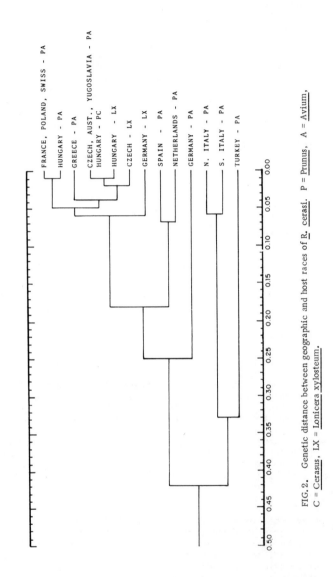

FIG. 2.   Genetic distance between geographic and host races of R. cerasi.   P = Prunus,   A = Avium,
C = Cerasus, LX = Lonicera xylosteum.

ACKNOWLEDGEMENTS

I would like to express my gratitude to Ms. Mary Jane Andrews and
Ms. Cynthia D. McWhorter for their technical assistance. I also
thank the many North American and European colleagues whose excellent cooperation in supplying samples for electrophoresis made
this study possible. This work was supported by a grant from the
National Institutes of General Medical Sciences, GM 15769.

REFERENCES

[1]   ANONYMOUS, Cooperative Econ. Insect Rep. 23, 155 (1973).

[1]   BOLLER, E. F., BUSH, G. L.   The population biology of the European
        cherry fruit fly, Rhagoletis cerasi L. (Diptera: Tephritidae)
        I. Evidence for genetic variation based on physiological parameters
        and hybridization experiments. Ent. exp. appl. (1974) In press.

[3]   BUSH, G. L., HUETTEL, R. N.   Starch gel electrophoresis of tephritid
        proteins. I.B.P. Working Group on Fruit Flies. Population Genetics
        Project Phase I. (1972).

[4]   BUSH, G.L. Population and ecological genetics of fruit flies.
        In I.B.P. Synthesis Report on Fruit Flies (1973). In press.

[5]   C.B.C., D.M.B.   The screw-worm strikes back. Nature, Lond. 242
        (1973)   493-494.

[6]   EWING, A. W., MANNING, A.   The evolution and genetics of insect
        behavior. Ann. Rev. Ent. 12 (1967) 471-494.

[7]   FLETCHER, L. W., O'GRADY, Jr. J.J., CLABORN, H.V., GRAHAM,O.H.
        A pheromone from male screw-worm flies. J. Econ. Ent. 59 (1966)
        142-143.

[8]   FLETCHER, L. W., CLABORN, H. V., TURNER, J. P.,LOPEZ, E.
        Differences in response of two strains of screw-worm flies to the
        male pheromone. J. Econ. Ent. 61 (1968) 1386-1388.

[9]   GEER, B. W., GREEN, M. M.   Genotype, phenotype and mating behavior
        of Drosophila melanogaster. Amer. Nat. 66 (1962)175-181.

[10]   HIGHTOWER, B. G. ADAMS, A. L., ALLEY, D. A. Dispersal of released
        irradiated laboratory-reared screw-worm flies. J. Econ. Ent. 58
        (1965)   373-374.

[11]   HIGHTOWER, B. G., O'GRADY,Jr., J.J., GARCIA, J.J. Ovipositional
        behavior of wild-type and laboratory-adapted strains of screw-worm
        flies. Envirn. Ent. 1 (1972) 227-229.

[12]   LEWONTIN, R.C., HUBBY, J.L.   A molecular approach to the study
        of genetic heterozygosity in natural populations. II. Amount of
        variation and degree of heterozygosity in natural populations of
        Drosophila pseudoobscura. Genetics 54 (1966) 595-609.

[13]   SELANDER, R.K., KAUFMAN,D.W. Genetic variability and strategies
        of adaptation in animals. Proc. Nat. Acad. Sci. USA 70 (1973)
        1875-1877.

[14]   SOKAL, R.R., SNEATH, P.H.A.   Principles of numerical taxonomy,
        W. H. Freeman. San Francisco. (1963).

QUALITY IN MASS-PRODUCED INSECTS

Definition and evaluation

D.L. CHAMBERS
Insect Attractants, Behavior and Basic Biology Research Lab.,
USDA, ARS,
Gainesville, Florida,
United States of America

Abstract

QUALITY IN MASS-PRODUCED INSECTS: DEFINITION AND EVALUATION.

The insect that is mass-produced and released in a control programme is in effect a biological bullet, a self-guided missile designed to deliver a beneficial effect against a pest insect. The ability of the released insect to achieve this objective may be influenced in many ways. The control of the quality of mass-produced insects must include an understanding of the behavioural components critical to their success and an evaluation of their performance based upon these behavioural components. The paper discusses some of the principles and techniques being used and developed to study behavioural performance and quality. Included are discussions of tests of: vigour, irritability, activity, sound production, response thresholds, reproductive preference and drive, biotic potential, and others.

1. INTRODUCTION

This paper is on the subject of insect behaviour as it relates to the control of fruit flies by the sterile release method and on the quality of mass-produced insects. The interest in this subject (it is a single subject) is burgeoning: a workshop was recently held at a Branch meeting of the Entomological Society of America; my agency, the Agricultural Research Service, will soon hold a pertinent workshop in Gainesville; and most of the recent panel meetings of OILB, IBP, and IAEA have applied considerable emphasis to or concentrated exclusively on this subject. Apparently entomologists have become aware that some of the problems encountered in pest control through release of mass-produced insects, entomophagous or sterilized, may occur because of behavioural deficiencies resulting from lack of control of quality.

We intend to make the study of principles and practices in behavioural quality measurement and management one of the major programmes at the Insect Attractants, Behavior and Basic Biology Research Laboratory of the Agricultural Research Service, USDA. Thus, I intend, in these few pages, to outline some of the philosophies and practices we will be using with a number of selected species. Among these is the Caribbean fruit fly, Anastrepha suspensa (Loew).

Before embarking on my discussion, I would like to refer the reader to the review of the subject of behaviour of mass-reared insects by Dr. Ernst Boller (1972). Because of his thoroughness, there is much I need not reiterate here.

2.   DEFINITIONS

Quality can be defined as the relative degree of excellence in some trait or skill that a thing possesses. Thus, when the term quality is applied to insects mass-produced for release in research or control programmes, it refers to the ability of the released insects to perform their function and to perform relative to some standard. The standard is generally taken to be the insect itself in either its native or untreated condition (i.e. not irradiated, not dyed, not selected by strain, etc.); the function is described by the nature of the control programme being imposed, as, for example, inducing sterility or mortality in the target population. The ability to perform the function refers herein to behavioural parameters (i.e. adaptedness, motility, and reproductive success), which are, ultimately, reflections of the genetic, physiological, and biochemical state of the insect.

Because ability is tied to function in the analysis of quality, "naturalness" (lack of deviation from the standard) is not necessarily the best measure of high quality. (Performance exceeding the standard could be desirable.) Conversely, selection of one or a few function-related traits as measures of overall quality can be dangerously misleading if, unknown to the scientist, other vital traits are lacking. An important challenge is to identify and rank the importance of the behavioural traits that contribute to the required function of released insects.

3.   STANDARDS

The entomologist responsible for control of quality in insect mass-production is, then, faced with two difficult decisions: (1) what is the most meaningful standard of reference; and (2) what are the behavioural traits that must be evaluated and how can they be measured? Ideally, one measures the performance of the expected function of the insects themselves and makes these decisions on the basis of direct information. These measurements may be feasible in control programmes where it is biologically possible to monitor results concurrently with releases, as in sterile male releases against mosquitoes where egg rafts can be collected and the level of sterility induced in the native population thus measured directly (Weidhaas et al. (1972)). Similarly, the percentage of parasitism occurring in a pest insect population may be monitored while parasite releases are underway. In these cases, the standard is the target insect population, the function is inducing suppression, and the measurement is the degree of success.

However, reliance upon the philosophy that "success is the best measure" is unacceptably risky in programmes such as sterile releases for fruit fly control where direct measurement of behavioural ability (inducing sterility) cannot be readily measured (here I mean large-scale tests or control programmes; egg fertility in field cages can be determined). In such programmes the usual criterion of success is population suppression or eradication, a result delayed by at least one reproductive cycle from the releases. Thus, by the time you know you are failing, based on population data, it may be too late to make corrections.

Hence, we must often develop standards and measurements for indirect monitoring, assuming thereby that we can select behavioural traits repre-

sentative of ability to achieve the function and that we can equate ability
in these traits to the likelihood of achieving the objective. (The risk in
these assumptions may also be large.)

Often there is little choice in the selection of standards. The released sterile
fruit fly most desirably retains those behavioural traits that are necessary for
competition with its native counterpart. Thus, the native fly is the most appropriate standard. However, it is exceptionally difficult to design and conduct tests
that adequately compare stock (mass-reared) and native insects without bias
in favour of one or the other: quantitative field data for specific behavioural
traits are difficult to obtain and interpret; conversely, confinement of native
flies in test cages necessarily alters their behaviour. Because of these
difficulties, we are often forced to use an internal standard. For
example, in a test of radiation effects we may compare an unirradiated
stock fly with an irradiated stock fly. Such internal standards are not
necessarily a poor compromise as long as the objective of the comparison
is measurement of change in stock quality and no immediate reference to
the native fly is assumed or intended. Moreover, acceptance of this
compromise allows one to proceed with selection and use of a number of
techniques for studying comparative behaviour in the laboratory and in
cages. From these studies, sufficient insight may develop to permit
selection and interpretation of one or of a few laboratory and field tests
wherein the native fly population provides a still more meaningful standard.
However, laboratory comparisons of stock and native flies should be
conducted under conditions that extract maximum performance from native
flies, not the reverse (which is often the case).

4.   MEASUREMENTS

The second decision I mentioned referred to the selection and measurement of skills for evaluating fly quality. I wish to emphasize very strongly
that when one devises, applies, and interprets laboratory or field cage
behaviour one must include the results of studies of the behaviour of native
flies in the field. Such studies must be of sufficient depth to indicate those
behavioural components that will be vital to the success of the released
insect and thus most meaningful for measurement. That is not to say that
all natural behavioural components will be necessary to the achievement
of the objective for which the insects are released or, conversely, that
only tests of vital functions are meaningful in measuring changes in stock
quality. Adequate knowledge is required of both natural behaviour and
programme function so sufficient priority and knowledgeable attention can
be given to control of significant parameters of quality.

The principal desired function of released sterile flies is mating and
this behavioural trait is often selected for measurement; indeed, it is
sometimes the sole measurement. To clarify my point I present the data
in Table I which appeared in a recent publication by Keiser et al. (1973)
concerning fruit fly mating behaviour.

It is quite clear that treatments B, C and D had no effect on mating
success as compared with the control (A). However, had these data been
developed as part of an assessment of fly quality (which is not the case)
it would be dangerous to assume that this measurement would be valid
for overall performance in a release programme because the treatments

TABLE I. EFFECT OF TREATMENT ON MATING SUCCESS

Treatment	Oriental fruit flies		Melon flies	
	No. eggs deposited	Hatch (%)	No. eggs deposited	Hatch (%)
A	249	62	180	94
B	581	66	360	93
C	332	78	120	93
D	346	67	240	94

consisted of total removal of one (C, D) or both (B) wings from the male. This is a deliberate overstatement of my point, and I use data gathered in a test that was not conducted in a manner quite like that of the usual mating test, the ratio test. Nevertheless, the analogy is close enough to validate my position, which is that ratio tests or similar mating tests, though assumed to give an overall view of performance of flies prepared for sterile release, are in small cage conditions actually only measurements of copulation. The fact is that behavioural trait probably has the lowest threshold of any mating trait, is one of the hardest to affect, and is, thus, one of the least sensitive to measurement. It is the prime and final trait in a chain of behavioural traits that could be weakened in earlier links and not be detected in a test of terminal function.

I believe that it is necessary to routinely evaluate the performance of flies produced and treated for release in each of the three major behavioural parameters that are functionally necessary — adaptedness, motility, and reproductive success. I further believe that selection of appropriate tests of these parameters can only follow (1) a fairly thorough study of the behaviour of the fly, (2) determination (perhaps subjective) of potential key behaviour traits, and at least preliminary estimates of which traits are subject to alteration through treatment or selection, (3) development of techniques for measuring those traits, and (4) acceptance of performance levels for those traits.

Measuring adaptedness

The meaning I intend for the parameter adaptedness includes such concepts as biological fitness, behavioural conformity, and genetic similarity. It is a more general category than motility or reproductive success and covers many possible behavioural traits, some of which are important to these other two categories. Thus, production or perception of pheromones may be important in reproduction, and orientation to appropriate hosts may be important in motility (distribution). However, there are some general tests that can identify changes in adaptedness and would thus signal that specific behavioural parameters need to be tested.

Dr. Bush discusses in these panel proceedings one of the most prominent and useful of these — survey of genetic content by enzyme analysis. This is a most useful tool for testing the highly important aspect of genetic hetero-geneity and divergence, as Dr. Bush has demonstrated.

Other physiological tests can alert one to changes in adaptedness. Some of these are: tests for changes in pesticide resistance; changes in metabolic function, such as CO_2 output, nutritional needs, or reserve content; changes in tolerance to temperature, irradiation, or other physical factors; changes in fertility, fecundity, longevity, or population stress tolerance; and changes in biological conformity such as rhythmicity, mating behaviour, host specificity, or chemical and physical cue production or recognition. Many others could also be proposed. The article by Langley (1970) indicates certain physiological parameters that could indicate adaptive shifts in reared populations.

Some of the techniques for studying insect behaviour that we are develop-ing and evaluating in our laboratory are useful for measuring adaptedness. Dr. Milton Huettel is applying the technique of enzyme analysis described by Dr. Bush to problems of variations in laboratory stocks and native strains of several insect species including Anastrepha suspensa. Also, Dr. William K. Turner, an agricultural engineer on our staff, is developing sensitive techniques for monitoring respiratory functions of insects for the study of insect response to radiation (Turner and Charity (1971), Turner et al. (1973)). These procedures should have ready application to the deter-mination of adaptive shifts in insect behaviour.

Dr. J.C. Webb, collaborating with Dr. John Sharp and with me, has been studying sound production and perception in A. suspensa (and other insects); Dr. Webb has developed an exceptional array of equipment for the study of insect sound, including a large anechoic chamber with con-trolled climatic conditions and electronic equipment for recording, analysing, and reproducing sounds. Figure 1 displays the sound prints

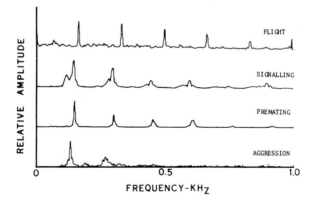

FIG.1.   Sound frequency signatures of male A. suspensa displaying the following behavioural traits: flight; signalling; premating (after mounting the female but before copulation); and aggression (attempted territorial competition).

TABLE II. ANALYSIS OF FLIGHT SOUND FREQUENCIES OF
Anastrepha suspensa

Age (days)	Treatment[a]	Power output (dB) in first fundamental	Power output (dB) in harmonics	No. of harmonics
1	A	10.4	40.1	7
	B	15.01	56.1	8
2	A	11.4	38.7	7
	B	13.0	70.5	11
3	A	8.7	45.3	8
	B	16.0	53.0	9

[a] A = normal flies; B = flies detrimentally affected in shipment.

characteristic of four behavioural activities of A. suspensa — flight, male
signalling, premating, and aggressive behaviour (Webb (1973)). We believe
it will be possible to monitor the sounds produced by a sample of a population reared and treated for release and thereby to determine both the
quantity and quality of the activity of that population. In other words, the
quality and relative distribution and rhythmicity of behaviour such as
resting, flying, aggression, mating, calling, etc., can be determined by
evaluation of the sounds produced.

    Table II shows the flight sound analysis of two batches of A. suspensa,
one considered to be normal and one suspected of having suffered detrimentally from shipping conditions. (We gratefully acknowledge the provision
of test insects by Dr. Richard Baranowski, University of Florida,
Homestead, Florida).

    Note that the presumably abnormal flies not only produced more sound
power in the first fundamental frequency but also more in both number of
harmonics and power in harmonics. We tentatively interpret this as the
result of irregular and/or unsynchronized wing movements. Other examples
of sound analysis are discussed later.

Measuring motility

    The ability of the released sterile fly to vigorously insert itself into
the habitat of the native pest insect has always been recognized as of prime
importance. Stock flies subjected to treatments such as irradiation, dyes,
packaging, and shipping have been noted to be relatively sedentary. A number
of tests of motive behaviour have therefore been developed. They can be
conveniently divided into two categories: (1) measurements of capability
(vigour, strength, velocity, endurance, etc.); and (2) measurements of
propensity (willingness, drive, threshold, etc.).

    Research to date has concentrated primarily on flight capability.
A number of us present at this panel have conducted flight mill studies with
fruit flies, and field release-recapture data are frequently reported.

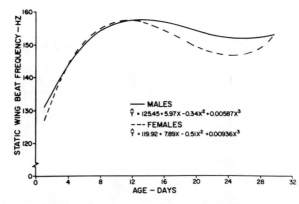

FIG.2. Relationships of fly age and frequencies of wing beats of male and female A. suspensa suspended without tarsal contact over a microphone, thus wing movement without lateral motion (static flight).

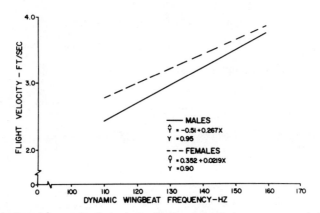

FIG.3. Flight velocity and frequencies of wing beats of male and female A. suspensa suspended from a flight mill rotor and passing over a microphone during lateral flight (dynamic flight).

I believe such tests are extremely important to quality testing. The flight mill system at the Gainesville laboratory will be used to evaluate the flight capability in A. suspensa; at present it is occupied in tests with other insects. This system, in its rotor design, is essentially like that of Schoenleber et al. (1970), though the design was suggested to me independently by an instrument maker in Honolulu. Dr. Boller also uses a similar system. In our flight mill, rotation of the arm interrupts a light beam to a photo-transistor, and this signal is transduced via an event recorder onto a tape recording that is analysed by a computer. However, a digital or strip chart recording can also give valuable data.

Another measurement of motive ability is wingbeat frequency. Figure 2 shows the wingbeat frequencies of male and female A. suspensa as a function of age. The data were developed by Dr. Webb (1973) by sound analysis.

FIG.4. Actographic equipment for measuring activity patterns and vigour: above - insect cages suspended in a frame on rubber bands; lower left – vibration detector mounted on top of a single cage; lower right – circuitry and recording equipment for transducing vibrations onto strip-chart recordings.

Frequency was similar for both sexes and increased until the flies were about 12 days old, then it decreased slightly until wing fraying evidently produced further increases. Sharp (1972) showed how stroboscopic measurement of wingbeat frequency can indicate irradiation effects. In our more recent tests, sound analysis coupled with the flight mill allowed simultaneous measurement of wingbeat frequency and flight velocity and showed the linear relationship (Webb (1973)) (Fig.3). I would also like to suggest that inexpensive and uncomplicated tests that indicate motive capability can be devised. For example, rate or frequency of flight across a darkened room to a lighted window can provide a measure of flight ability.

It is my opinion that flight propensity may be a behavioural trait that is equally or more important than flight capability in the success of a sterile release programme. The initial distribution of the released flies can cover the infested areas, but the flies must aggressively search out the proper habitats and locate mates. In fact, low motive propensity has been suspected as a reason for lack of success in several fly release programmes. A test for motive propensity is therefore important, and I refer the reader to two simple methods of obtaining this measurement. Schroeder et al. (1973b) described an inexpensive technique for measuring flight propensity. With this unit we have shown that irradiated stock Mediterranean fruit flies were approximately 50% less prone to flight than unirradiated or native flies. In addition, we also use at our laboratory an actographic device developed by Dr. Norman Leppla, based on his original design (Leppla and Spangler (1971)). Banks of screen cages suspended on rubber bands are fitted individually with silica chip transducers (microphone-type pickup devices) that are inexpensive detectors of vibrations. The movements of insects against the cage walls are detected and recorded (Fig.4). Since both amplitude and frequency are recorded, the system provides measurements of the frequency and vigour of movement as well as indicating circadian characteristics. (As mentioned earlier, sound analysis can also indicate the kind and quality of movement in a cage of flies.)

Much valuable information concerning flight behaviour in the field can be obtained, and many publications describe methods of conducting such field studies. A good reference source is the article by Bateman (1972). Also, Dr. Prokopy (1975) has presented the results of novel and potentially useful new techniques for field studies of fruit fly movement. At the Attractants and Behavior Laboratory, we have an active programme of developing and using techniques for studying insect movement in the field (at present, insects other than fruit flies). Thus, my concentration in this brief paper on our laboratory data should not imply that we are neglecting the important field studies.

Measuring reproductive success

Boller (1972) presented a thorough discussion of premating and post-mating mechanisms that may result in reproductive failure in released insects. I will use these same divisions to discuss some of the behavioural traits that are probably needed in released sterile fruit flies and some of the tests used to evaluate performance of specific behavioural traits. The typical ratio test I believe spans both pre- and postmating parameters

(without identifying wherein failure might lie but serving as a screen for
both areas). See Fried (1971) for a comprehensive discussion of ratio test
analysis.

In the first category, no mating occurs.

A. Potential mates do not meet
    a. spatial dislocation
    b. altered periodicity of mating activity
    c. communication failure
    d. disconformity of physiological state or age

B. Potential mates meet but do not mate
    a. communication failure
    b. physiological disconformity
    c. competitive displacement

In the second category, a sterile mating takes place, but the mated
female is not removed from the reproducing population.

C. Copulation occurs but subsequent mating with a fertile male results
    in female fertility.

D. Copulation occurs but the sterile mating does not induce rejection of
    subsequent males in the native female.

Testing premating factors

Ethological isolation is probably best observed in the field (Prokopy
(1968)) or in large field cages (Katiyar and Ramirez (1970); Holbrook and
Fujimoto (1970); Ohinata et al. (1971)). Thus, it has been possible to
observe altered periodicity, size and strain preference, site preference
and competitive displacement. Spatial dislocation can be discovered
through careful field observations. Disconformity of physiological state
or age might be suspected when cage tests indicate mating success but
field data do not; then survival factors (inadequate longevity), early release,
or retarded maturation through treatment or handling would be suspected
(see Haniotakis (1973)). However, laboratory tests of behavioural conformity
are certainly possible. Schroeder et al. (1973a) demonstrated in uniquely
simple laboratory tests that gamma irradiation of D. cucurbitae as pupae
delayed sexual maturation 24 hours and reduced mating intensity, but large
cage tests were required to determine that time of mating was not altered.

Communication failure is an aspect of behaviour that interests us at
our laboratory. Dr. Webb, Dr. Sharp and I have developed preliminary data
on the signalling sound produced by male A. suspensa and some of the effects
of gamma irradiation on its physical and biological characteristics. Note
in Fig.5 that the two-pulsed sound is blurred and distorted following irradiation with 10 kR. Also, a characteristic component is delayed. Table III
shows the responses of normal stock flies and 'abnormal' stock flies
(believed exposed to excessive temperature in shipment) to combinations
of reproduced male calling sound and extracts of male pheromone
(graciously supplied by Dr. J.L. Nation, University of Florida, Department
of Entomology and Nematology). In all comparisons, the communication

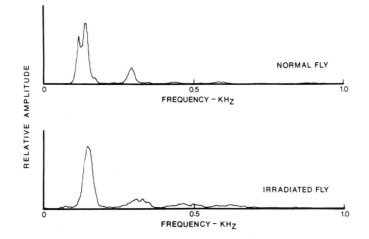

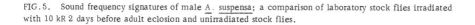

FIG.5.   Sound frequency signatures of male A. suspensa: a comparison of laboratory stock flies irradiated
with 10 kR 2 days before adult eclosion and unirradiated stock flies.

TABLE III.   RESPONSES OF Anastrepha suspensa TO SIGNALLING
SOUNDS, PHEROMONE EXTRACTS AND LIVE MALES

Treatment[a]	Average No. of flies responding	
	Normal	Abnormal[b]
Male to male	14.4	6.7
Male to sound	13.9	4.3
Male to extract	13.0	7.1
Male to sound + extract	13.4	2.0
Female to male	19.1	11.3
Female to sound	12.9	6.0
Female to extract	18.9	11.3
Female to sound + extract	19.8	8.3

[a] Sixteen replications of normal flies, three of abnormal flies.
[b] Flies exposed to detrimental conditions during shipping

response of the abnormal flies was decreased. I repeat that these data are preliminary and inconclusive, but they indicate a line of research we find most interesting.

Testing postmating factors

In the interest of brevity I present here a partial list of references to studies of the factors affecting postmating success, with brief descriptions of the findings. The procedure will indicate some of the ways postmating success can be measured and will indicate the importance such factors may have in achieving the objective of sterile releases.

A. Anwar et al. (1971); histological study showed that Mediterranean fruit flies irradiated as adults contained more sperm than those irradiated as pupae.

B. Barton Browne (1957); mated female Dacus tryoni repel subsequent mating attempts by males.

C. Tzanakakis et al. (1968); a substance transmitted in the semen may inhibit subsequent female mating in D. oleae.

D. Nakagawa et al. (1971); male Mediterranean fruit flies are polygamous but females remated less frequently, with receptivity inversely correlated with the amount of sperm remaining in the spermathecae.

E. Ohinata et al. (1971); male Mediterranean fruit flies irradiated as adults transferred more sperm than those irradiated as pupae.

F. Katiyar and Ramirez (1970); female Mediterranean fruit flies mated sequentially with sterilized and fertile males demonstrated that sperm mixing occurs and that a fertile mating after a sterile mating altered female fertility more than the reverse sequence.

5. CONCLUSIONS

A. Quality is a value judgement, not a biological constant.

B. Function is the name of the game, not productiveness and not naturalness.

C. 'Unnaturalness' should be a warning signal, not a criterion for rejection.

D. The native target insect is the best standard of reference, but an internal standard is acceptable if its limitations are acknowledged.

E. Performance levels are needed, but according to behavioural traits and function, they may acceptably be substandard, standard, or above standard.

F. Some behavioural traits are obviously needed, but they may not be the best traits to measure and should never be the only traits measured.

G. Know your insect in the field and laboratory, know your programme function, know your programme limitations, and give quality a proportionate priority.

H. Test at least:
   a. Adaptedness — do you still have an insect sufficiently like the target insect to achieve its function?
   b. Motility — can and will your insect get to the target?
   c. Reproductive success — can the flies you produce suppress native fly fertility with acceptable efficiency?

REFERENCES

ANWAR, MAHMOOD, CHAMBERS, D.L., OHINATA, K., KOBAYASHI, R.M. (1971), Radiation sterilization of the Mediterranean fruit fly (Diptera: Tephritidae): Comparison of spermatogenesis in flies treated as pupae or adults, Ann. Entomol. Soc. Am. 64(3), p.627.

BARTON BROWNE, L. (1957), An investigation of the low frequency of mating of the Queensland fruit fly Strumeta tryoni Frogg, Aust. J. Zool. 5, p.159.

BATEMAN, M.A. (1972), "The ecology of fruit flies", Annual Review of Entomology, Vol.17 (R.F. SMITH, MITTLER, T.R., SMITH, C.N., Eds), Annual Reviews, Inc., Palo Alto, p.493.

BOLLER, E. (1972), Behavioral aspects of mass-rearing of insects, Entomophaga 17(1), p.9.

FRIED, M. (1971), Determination of sterile-insect competitiveness, J. Econ. Entomol. 64(4), p.869.

HANIOTAKIS, G.E. (1973), Sexual competitiveness of metepa-sterilized males of Dacus oleae, Environ. Entomol. 2(5), p.731.

HOLBROOK, F.R., FUJIMOTO, M.S. (1970), Mating competitiveness of unirradiated and irradiated Mediterranean fruit flies, J. Econ. Entomol. 63(4), p.1175.

KATIYAR, K.P., RAMIREZ, E. (1970), "Some effects of gamma radiation on the sexual vigour of Ceratitis capitata (Wiedemann)", Sterile-male Technique for Control of Fruit Flies (Proc. Panel Vienna, 1969), IAEA, Vienna, p.83.

KEISER, I., KOBAYASHI, R.M., CHAMBERS, D.L., SCHNEIDER, E. (1973), Relation of sexual dimorphism in the wings, potential stridulation, and illumination to mating of oriental fruit flies, melon flies, and Mediterranean fruit flies in Hawaii, Ann. Entomol. Soc. Am. 66(5), p.937.

LANGLEY, P.A. (1970), "Physiology of the Mediterranean fruit fly in relation to the sterile-male technique", Sterile-male Technique for Control of Fruit Flies (Proc. Panel Vienna, 1969), IAEA, Vienna, p.25.

LEPPLA, N.C., SPANGLER, H.G. (1971), A flight-cage actograph for recording circadian periodicity of pink bollworm moths, Ann. Entomol. Soc. Am. 64(6), p.1431.

NAKAGAWA, S., FARIAS, G.J., SUDA, D., CUNNINGHAM, R.T, CHAMBERS, D.L. (1971), Reproduction of the Mediterranean fruit fly: frequency of mating in the laboratory, Ann. Entomol. Soc. Am. 64(4), p.949.

OHINATA, K., CHAMBERS, D.L., FUJIMOTO, M., KASHIWAI, S., MIYABARA, R. (1971), Sterilization of the Mediterranean fruit fly by irradiation: comparative mating effectiveness of treated pupae and adults, J. Econ. Entomol. 64(4), p.781.

PROKOPY, R.J. (1968), Visual responses of apple maggot flies, Rhagoletis pomonella: orchard studies, Entomologia Exp. Appl. 11, p.403.

PROKOPY, R.J. (1975), these Proceedings.

SCHOENLEBER, L.G., WHITE, L.D., BUTT, B.A. (1970), Flight mill system for studying insect behaviour, USDA-ARS 42-164, 7 pp.

SCHROEDER, W.J., CHAMBERS, D.L., MIYABARA, R.Y. (1973a), Reproduction of the melon fly: mating activity and mating compatibility of flies treated to function in sterile-release programs, J. Econ. Entomol. 66(3), p.661.

SCHROEDER, W.J., CHAMBERS, D.L., MIYABARA, R.Y. (1973b), Mediterranean fruit fly: propensity to flight of sterilized flies, J. Econ. Entomol. 66(6), p.1261.

SHARP, J.L (1972), Effects of increasing dosages of gamma irradiation on wingbeat frequencies of Dacus dorsalis Hendel males and females at different age levels, Proc. Hawaii Entomol. Soc. 21(2), p.257.

TURNER, W.K , CHARITY, L.F. (1971), Determining response of insects to radiation by continuous monitoring of their carbon dioxide output, Ann. Entomol. Soc. Am. 64(2), p.419.

TURNER, W.K., CHARITY, L.F., SODERHOLM, L.H. (1973), Use of automation and computer analysis in a study of insect response to radiation, Trans. Am. Soc. Agric. Engineers 16(5), p.956.

TZANAKAKIS, M.E., TSITSIPIS, J.A., ECONOMOPOULOS, A.P. (1968), Frequency of mating in females of the olive fruit fly under laboratory conditions, J. Econ. Entomol. 61(5), p.1309.

WEBB, J.C. (1973), Analysis and identification of specialized sounds possibly used by the Caribbean fruit fly Anastrepha suspensa (Loew) for communication purposes, Doctoral dissertation, University of Tennessee, Knoxville (Univ. Microfilms, Inc., Ann Arbor, Mich.).

WEIDHAAS, D.E., LABRECQUE, G.C., LOFGREN, C.S., SCHMIDT, C.H. (1972), Insect sterility in population dynamics research, Bull. World Health Organ. 47, p.309.

PROGRESS OF THE PILOT TEST AT LANAI AGAINST MEDITERRANEAN FRUIT FLIES AND MELON FLIES*

E.J. HARRIS, R.T. CUNNINGHAM,
N. TANAKA, K. OHINATA
Hawaiian Fruit Flies Lab.,
USDA, ARS,
Honolulu, Hawaii,
United States of America

Abstract

PROGRESS OF THE PILOT TEST AT LANAI AGAINST MEDITERRANEAN FRUIT FLIES AND MELON FLIES.
A pilot test programme is underway on Lanai, State of Hawaii. The objective is the suppression of the Mediterranean fruit fly with sterile medfly releases and the melon fly with male annihilation treatments. Plans have been laid, and preliminary testing has begun to determine the most effective approaches.

INTRODUCTION

The presence of insect pests such as the melon fly, Dacus cucurbitae Coquillett, the oriental fruit fly, D. dorsalis Hendel, and the Mediterranean fruit fly, Ceratitis capitata (Wiedemann), in the State of Hawaii presents a serious obstacle to the development of diversified agriculture. Not only do these flies cause direct injury to fruits and vegetables; they also make it necessary for the state and federal governments to maintain a large quarantine programme to minimize the risk of accidentally introducing the flies into the mainland United States. Indeed, fruit flies have been estimated to cost Hawaii US $ 52 573 200[1] per year by their existence in the state. Therefore, since state and federal officials would like to eradicate or neutralize the impact of these flies on the economy of Hawaii, the Hawaiian Fruit Flies Laboratory of Agricultural Research Service, US Department of Agriculture, has received special funds to study methods for large-area treatment and to determine the feasibility of eradiacting fruit flies from Hawaii.

LOCATION OF TEST AND OBJECTIVES

The 381-km^2 island of Lanai (Fig. 1) was selected for the pilot test because its size and location, 119 km from Honolulu, are appropriate to the resources and equipment that are available to carry out the experiment.

---

* Mention of a pesticide or a commercial product or company in this paper does not constitute a recommendation or an endorsement by the USDA.

[1] An Appraisal of the Importance of Fruit Flies to Hawaii's Economy, prepared by the staff of the Division of Plant Industry, State of Hawaii, Department of Agriculture Sept. 29, 1966.

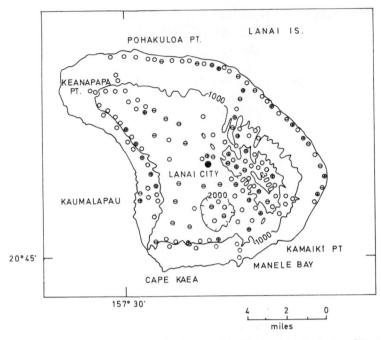

FIG. 1.   Map of Lanai showing the topographic features of the island and the distribution of Hawaiian tephritid fruit flies. Elevation in feet (ft ÷ 3.28 = metres). Distance in miles (miles × 1.61 = kilometres). $\oplus$ = medfly; $\ominus$ = melon fly; $\bigcirc$ = oriental fruit fly.

In addition, the island has appreciable populations of all three species of flies. Since the islands of Molokai and Maui are located 13 - 16 km north and east northeast, respectively, from Lanai, we do not expect complete eradication of the fruit flies from Lanai. They will be able to migrate on the trade winds from the neighbouring islands.

Lanai is usually a dry island (about half receives less than 51 cm of rain per year), and wild host plants for Tephritidae are found for the most part in the 31-km² area that has higher rainfall, the central high peak (305 - 610-m elevation). However, cultivated and wild hosts also occur in the central village (1.29-km² area, 2500 population) and in a few other small areas. The cultivated hosts are probably chiefly important in helping to carry the medfly over between guava seasons (though there is never a time without guava) and in providing adult food and shelter.

The programme will consist of six essentially separate operations: a fruit fly survey; the development of methods of treating large areas; two programmes, one involving the release of sterile Mediterranean fruit flies (medfly) and one involving treatment of melon flies by male annihilation and by releases of sterile melon flies; studies of the environmental impact of the programme; and the development of systems of evaluating the programme.

The survey programme is now in progress and will continue throughout the experiment (it consists chiefly in monitoring the catches in approximately 225 triplets of plastic traps (675 total), one each for the medfly, melon fly, and oriental fruit fly, distributed throughout all the accessible areas of the island) because trap catches are the chief tool in determining the pattern of distribution of both released and native flies. For example, recaptures of released dye-marked medflies and melon flies are used to delineate dispersal, survival, and overflooding patterns, which suggest modifactions in treatment. Also, the information from the 675 traps is being supplemented by surveys for host biomass indices of fruit infestation and evaluation of the gravidity and insemination of females captured in liquid-bait traps. Thus, the survey and also the other operations support the effort to demonstrate suppression.

As the programme indicates, we are giving top priority to the development of the sterile-insect technique for control of the medfly; next most important is the development of the male-annihilation technique and the release of sterile flies for control of the melon fly. We have already developed the male-annihilation technique for control of the oriental fruit fly sufficiently so that we are confident it will perform satisfactorily. Nevertheless, if the pilot test progresses well against the medfly and the melon fly and resources permit, we will enlarge the programme to include further development of the male-annihilation technique against the oriental fruit fly. In this event suppression of all three species of fruit flies on Lanai would be achieved.

RELEASES OF STERILE MEDITERRANEAN FRUIT FLIES

We began releases of sterile medflies in May 1973 at an average rate of 1 - 10 million pupae per week. At that time, the pupae were reared in Honolulu (irradiated with 10 krad 1 day before adult eclosion and dyed with Calco® Blue powder) and airmailed to Lanai in anaerobic poly bags (Tanaka et al., 1972). This modest production has now been increased to 15 to 20 million pupae per week. We handle some 4 - 5 million of the flies each week by bulk chilling and holding at 3 - 5°C, irradiating (as adults at 10 krad) in lots of 500 - 700 thousand, and packaging in paper bags (25 000 per bag) mixed with sphagnum moss, alfalfa hay, or oat straw (Schroeder et al. of this laboratory, unpublished data). The bagged flies are subsequently dropped on Lanai from a light airplane. Also, about 1 million flies are dropped as naked pupae or adults from a refrigerated dispensing machine supplied by the Animal and Plant Health Inspection Service (APHIS). The remainder of the flies, 10 - 15 million, are irradiated as pupae (in a nitrogen atmosphere) at 12 krad, shipped in poly bags, and released 2 - 4 times per week at ground sites from 1.892-litre plastic buckets that hold as many as 15 000 - 20 000 pupae per bucket. The 600 ground release sites are located throughout the 31-km^2 area of the island where breeding hosts are found. Thus, the aerial releases supplement the ground releases.

The new methods of handling, irradiating, and releasing the sterile medflies evolved from a series of preliminary tests made to compare the survival of flies released as fed adults with that of flies allowed to emerge as pupae in the field. The dyed irradiated pupae were placed in 5.44-kg paper bags (Nadel et al., 1967) at a rate of 1500 pupae per bag, and each bag was provided with a food and water source. Then the flies were held in ambient temperatures until most of the adults emerged. At this point, about 100 000 of these flies and 100 000 pupae (handled as described) were placed at each of 10 sites on Lanai. The resulting recovery from the adult releases was about 2 times the recovery from the pupal releases in the two trials. (However, because of logistical difficulties, most flies that we have released thus far have been irradiated as pupae.) Likewise, aerial drops of refrigerated adults or undyed naked pupae were compared with ground releases of pupae. The rates of recovery of pupae were similar; the rates for the refrigerated adults were much higher. Nevertheless, recoveries from all releases were low, and the methods need further refinement.

Initially, the dispersal and survival of adults emerging from the released pupae were low because of the atypical harsh, windy, and dry conditions of the island; however, survival improved when rainfall increased. Also, pupae in poly bags were injured and fewer adults emerged when temperatures rose above 25°C; this injury is reduced by placing ice blocks in the wood shavings placed around the pupae when they are sent by airmail from Honolulu to Lanai. Nevertheless, the system of shipping the pupae by mail still is supplemented by air freight shipment to minimize overheating.

We also investigated the use of aerial sprays of PIB-7 hydrolysed protein-technical malathion (4:1 ratio) applied 4 times at a rate of 215 g active malathion per hectare to prevent the seasonal August peak of native medflies (we also assessed the sprays as a potential spot treatment for control in the medfly-breeding area). Sterile medflies, native medflies, and oriental fruit flies were all killed by the treatments, and the seasonal medfly peak did not occur; in addition, populations of oriental fruit flies in the area were kept low. However, the sprays were only effective for 2 weeks, and they have been discontinued.

With the releases, we are currently achieving an overflooding ratio of 80:1 released to native males, and the overflooding ratio is increasing. However, after any one release, a number of traps catch only undyed flies (some are undoubtedly sterile flies that have lost dye). We expect to improve greatly on this ratio as we increase production, re-allocate release sites, and use the sterile medflies more judiciously.

Releases of sterile medflies will be continued at least through November 1974 so we can measure the incidence of medfly larvae in the peak-season guava crop.

MALE-ANNIHILATION TREATMENT OF MELON FLIES

On Lanai, the melon fly is distributed in clusters in close association with cultivated or wild breeding hosts. Thus, the fly is found in small gardens in the 1.29-km² village and in older pineapple fields where Momordica charantia, the principal breeding host, grows as a weed. The male-annihilation treatments will therefore be applied to areas that have Momordica. These areas were delineated in co-operative studies with the Citrus Insects Investigations Laboratory, Weslaco, Texas (Hart et al., 1971) in which infrared aerial colour photographs were made of 7290 hectares of pineapple fields. This remote sensing technique, a promising survey tool that we hope to use again, allowed us to identify 364.5 hectares of pineapple fields containing various-size patches of Momordica.

The tests of male annihilation against melon flies are at present planned to proceed as follows: In the first place, to attract melon flies out of the pineapple fields, we will use light aircraft to apply 4.47 kg per hectare of Thixcin-thickened sprays of cue-lure (Ohinata et al., 1971) (60% cue-lure + 10% Thixcin + 30% naled) to 72.5 km^2 around the perimeter of pineapple fields. Also, we will apply four swaths, 0.161 km apart, extending 1.61 km out from the border of the pineapple fields toward the sea. In addition, in the Keamoku Beach area on the north side of Lanai, three swaths will be applied extending from the shoreline inland, and aerial sprays of the formulation will be applied over the remaining inhospitable area of the island (ca. 207.2 km^2) at the rate of 140 g of mixture per km^2 at 3-week intervals (to ensure adequate coverage of the portion of the island that probably harbours only transient flies). Finally, in the inhabited areas, the villages, we will apply the mixtures to powerline poles, trees, hedges, and fence posts by squirting it from oil cans at a rate of 0.2 g per squirt per 15.25 m (148 or 174 squirts per hectare).

In the second place, 15.24-cm pieces of 24-ply string saturated with 70% cue-lure + 30% naled (0.9 of mixture per 15.24-cm string) will be applied to 608 hectares of post-harvest pineapple fields at a rate of 40 strings per hectare (91.5 m apart). However, we have since found that cigarette filter tips can be substituted for the strings. Also, we will treat 1012.5 - 1215 hectares of post-harvest pineapple fields by locating redwood stakes (1.27 cm × 91.44 cm) saturated with 4 - 6 g of 70% cue-lure + 30% naled around the perimeter of the 45.8-m beds at the rate of 12 stakes per hectare. Fresh stakes will be set out at 2-month intervals.

The rationale for the sequence is that the ease of application and good coverage obtainable with the thickened sprays will produce a more drastic population reduction than the strings or filter tips. Then once suppression is achieved, it can be maintained and enhanced by the longer-lasting strings or filter tips that take about twice as long to apply.

A preliminary treatment, a single aerial spray of cue-lure + naled thickened with Thixcin, was applied to a 30.49 km × 0.805 km border strip around one of the fields of pineapple. Immediately after the application, traps adjacent to this sprayed area showed a reduction in catch of 98% compared with the pretreatment level; after 2 weeks, the reduction was 88%. The spray was therefore judged sufficiently efficacious to use in the male-annihilation programme, and treatment intervals were set at 2 - 3 weeks. In another preliminary test, a 243-hectare Momordica-infested pineapple field was treated by distributing redwood stakes soaked at the tip with 5 g of cue-lure + naled at the rate of 12 stakes per hectare. No male melon flies were recovered from that field during the first 8 weeks post-treatment.

We have also been testing the equipment that will be used. For example, in co-operative studies, Agricultural Research Service engineers in Yakima, Washington, have developed a string cutter, a solid particle auger, and a viscous-lure spray pump that will be used to apply treatments from aircraft. The string cutter and auger were used to distribute male-annihilation treatments in 688.5 hectares of post-harvest pineapple fields that had Momordica infestations; the spray pump was used to apply spray to the uninhabited portions of the island that are not in pineapple. Also, in October 1973, we made further checks of equipment and methods. The liquid-lure dispenser malfunctioned and required extensive modification and redesign.

The treatments with 2.36-cm pieces of 24-ply string were not as effective as expected. Finally, heavy rainfall in December and January reduced the effectiveness of the treatments.

COMMENTS

The procedural difficulties with the treatments against the melon fly have led to the decision to hold the male-annihilation programme in abeyance until we finish the release programme against the medfly. Liquid male-annihilation treatments only may be made for compatibility with the releases of sterile melon flies to mop-up wild flies missed by the treatments.

Our general strategy requires that we first test the methods of treatment to determine their efficacy. We must also determine the best way to conduct the experiment on Lanai before we commit ourselves completely to any one approach. When the data are not conclusive, the approach we choose will be based on the simplest, least expensive, or most logical method. We will keep our options open until we are sure we are proceeding in the right direction. We make changes in our methods as needed and have been able to effect savings in the cost of materials and applications at the same time that we have improved our methods.

REFERENCES

HART, W.G., INGLE, S.J., DAVIS, M.R., MANGUM, C., HIGGINS, A., BOLING, J.C. (1971), Some uses of infrared aerial color photography in entomology, Proc. Third Biennial Workshop in Color Aerial Photography in the Plant 1971.

NADEL, D.J., MONRO, J., PELEG, B.A., FIGDOR, H.C.F. (1967), A method of releasing sterile Mediterranean fruit fly adults from aircraft, J. Econ. Entomol. 60, p. 889.

OHINATA, K., STEINER, L.F., CUNNINGHAM, R.T., Thixcin[®] E as an extender of poisoned male lures to control fruit flies in Hawaii, J. Econ. Entomol. 64, p. 1250.

TANAKA, N., OHINATA, K., CHAMBERS, D.L., OKAMOTO, R. (1972), Transporting pupae of the melon fly in polyethylene bags, J. Econ. Entomol. 65, p. 1727.

THE PROCIDA MEDFLY PILOT EXPERIMENT

Status of the medfly control
after two years of sterile-insect releases[*]

U. CIRIO
Laboratorio Applicazioni in Agricoltura,
Comitato Nazionale per l'Energia Nucleare,
Rome, Italy

Abstract

THE PROCIDA MEDFLY PILOT EXPERIMENT: STATUS OF THE MEDFLY CONTROL AFTER TWO YEARS
OF STERILE-INSECT RELEASES.
    The CNEN initiated in 1971 a four-year pilot experiment of medfly control by the sterile insect technique
on the island of Procida. The mass rearing and sterilization of the medfly, the logistic techniques, the
evaluation method and the control strategy used in 1973 are described. The medfly was effectively suppressed
by releasing a limited number of sterile flies and by other simple control methods.

INTRODUCTION

    During the past twenty years, in spite of extensive use of pesticides,
the losses and problems to the Italian fruit industry due to Ceratitis
capitata (Wiedemann) have increased. It was recently estimated in Italy
that the loss of income due to this species is about 45 million dollars a
year [1]. The urgent need for alternative methods to control medfly induced
the Laboratory for Application of Nuclear Techniques to Agriculture, CNEN,
under the auspices of the IAEA, to explore the possibilities of using the
sterile-insect technique (SIT) in the Parthenopean islands [2, 3]. These
first trials, conducted in collaboration with the Italian Ministry of Agriculture
and Forestry (MAF), EURATOM and IAEA, proved very promising. The
necessity, however, to have definitive, practical and economic data on the
application of this technique convinced CNEN to begin, in 1971, a four-year
pilot experiment characterized by ecological studies during and after the
two initial years of sterile fly releases. This experiment, located again
on the island of Procida, is carried out in collaboration with MAF and
EURATOM. The present paper describes the control strategy followed in
the 1973 releases and discusses the results obtained.

---

[*]Contribution 395 from the CNEN Agricultural Laboratory, Casaccia Nuclear Studies Centre,
S. Maria di Galeria, Rome, Italy.

FIG. 1.   Release cage used for sterile medfly adults on the island of Procida.

MATERIALS AND METHODS

1. Mass rearing and irradiation of medfly

The flies were reared at the Casaccia insectary using the technique reported by De Murtas and Cirio [4]. During the winter a Ceratitis wild population from Procida was colonized in the insectary. At irregular intervals during the year, many flies collected from fruits artificially infested in field cages at the insectary were introduced into the mass rearing. Studies on the modification of egg incubation, cheaper larval diets, and mechanization of rearing procedures were carried out or are still in progress. The length of the pupal stage was regulated by varying the temperature from 15 to 25°C to synchronize adult emergence. A few days before the end of the pupal period all insects were coloured with different fluorescent dyes according to the release areas. The sterilization of marked adult flies was accomplished at the ^{60}Co irradiation plant of the Agriculture Application Laboratory of CNEN — Casaccia Center — with a dose of 10 000 (±10%) rad. After irradiation, the sterile insects were transported during the night by car to Procida. During the first month, adult sterilization was accomplished by using paper bags as described by Cirio and De Murtas [5]. However, from June to October, following successful preliminary trials, the adult flies were irradiated in special 80 cm × 40 cm × 28 cm cloth cages, each containing about 100 000 insects that had been placed in the cages 3 days before their emergence (Fig.1). The cages were kept in air-conditioned cells at 25°C until the time of irradiation. The emerged adults were fed by wetting strips of paper towelling with water. The strips were hung in the cloth cages to increase the support surface for the flies. In addition to these cages, 50 paper bags (each containing approximately 10 000 insects) were sent to Procida. These extra flies were released in areas not easily accessible by car.

2. Distribution of sterile flies

The release of sterile adult flies on Procida began on 4 May 1973, later than planned in the control strategy. During the first month the paper bags with sterile insects were hung and ripped open mainly in the most favourable medfly over-wintering areas; these generally correspond with citrus cultivation. From June onwards, releases were carried out using small cars carrying two sterile insect cloth cages set upright and opened on the top. The releases were made on a weekly schedule along all roads of Procida, which practically are no more than 200 m from each other. Sterile marked insects were also released on Ischia and Monte di Procida to obtain information on Ceratitis movement. The numbers of sterile flies released are given in Table I. After each release the number of weak flies remaining on the bottom of the cages was estimated to calculate the real number of competitive insects released.

3. Evaluation of effectiveness of medfly control

In 1973, the methods of evaluating control effectiveness were simplified in order to reduce the field work and were restricted to the examination of fruit infestation and to trapping. On Procida, an accurate inspection of

TABLE I. DATA ON RELEASE OF STERILE MEDFLY IN THE
EXPERIMENTAL AND CONTROL AREAS[a]

MONTH	LOCALITY		
	Procida	M. di Procida[b]	Ischia
May	13 408 000	0	0
June	8 510 000	0	0
July	3 217 000	4 101 000	130 000
August	1 722 000	5 495 000	65 000
September	5 947 000	0	0
October	8 092 000	0	0
November	0	0	0
TOTAL	40 896 000	9 596 000	195 000

[a] Data refer to sterile adults that left the cages.
[b] Releases were made in collaboration with the Experimental Station of Italian Ministry of Agriculture &
Forestry, Entomological Section, Portici, Naples.

fruit selected in the 12 systematic sampling areas as well as in the control
areas was carried out every week according to the techniques described by
De Murtas et al. [2]. Attack was always indicated by the presence of
mature larvae in the fruit. On the island, further controls were made on
fruits imported from the mainland. All host fruits found in the stores and
suspected of being infested were gathered and put in special boxes to collect
the pupae. Indirect control of the effectiveness of the techniques was carried
out by trapping the flies in the Trimedlure bait traps always present on
Procida and in the control areas. The number and colour of the captured
flies were compared with those of adults caught in Capri, Ischia and
Monte di Procida.

4. Medfly control strategy

In 1973, the procedure for controlling Ceratitis on Procida was changed
according to the following considerations:

(a) The clearly successful 1972 Procida experiment [5];
(b) The lack of long-term eradication in the previous year's Procida
     programme (despite the continuous release of 85 million sterile adult
     insects from March to the end of October) due to the immigration of
     gravid females from the nearby areas [5];
(c) The ecological information collected on Procida and control areas
     since 1967 [5-13];
(d) The attempt to achieve the same results as in 1972 with a cheaper
     control programme using smaller numbers of sterile insects and
     simplifying the techniques [14].

PLANNING SEQUENCES

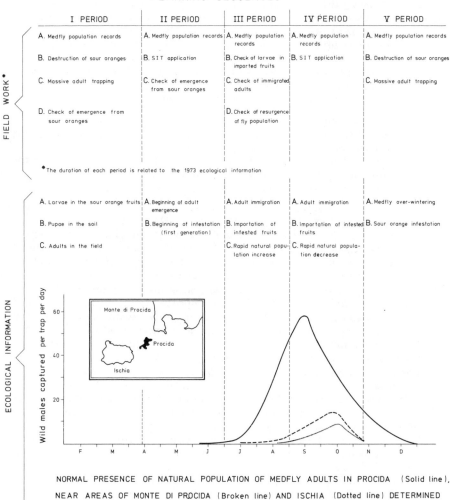

I PERIOD	II PERIOD	III PERIOD	IV PERIOD	V PERIOD
A. Medfly population records	A. Medfly population records	A. Medfly population records	A. Medfly population records	A. Medfly population records
B. Destruction of sour oranges	B. SIT application	B. Check of larvae in imported fruits	B. SIT application	B. Destruction of sour oranges
C. Massive adult trapping	C. Check of emergence from sour oranges	C. Check of immigrated adults		C. Massive adult trapping
D. Check of emergence from sour oranges		D. Check of resurgence of fly population		

* The duration of each period is related to the 1973 ecological information

A. Larvae in the sour orange fruits	A. Beginning of adult emergence	A. Adult immigration	A. Adult immigration	A. Medfly over-wintering
B. Pupae in the soil	B. Beginning of infestation (first generation)	B. Importation of infested fruits	B. Importation of infested fruits	B. Sour orange infestation
C. Adults in the field		C. Rapid natural population increase	C. Rapid natural population decrease	

NORMAL PRESENCE OF NATURAL POPULATION OF MEDFLY ADULTS IN PROCIDA (Solid line), NEAR AREAS OF MONTE DI PROCIDA (Broken line) AND ISCHIA (Dotted line) DETERMINED BY TRIMEDLURE BAITED TRAPS IN 1968

FIG. 2.   Strategy of medfly control suggested for 1973 experimental programme on the island of Procida.

The 1973 control programme (see diagram in Fig. 2) can be divided into a
sequence of five periods.

I.   The first period was devoted to reducing the over-wintering population
     by destroying the sour oranges and by intensive sticky trapping. This
     latter work began at the end of 1972 in order to eliminate the adult
     population and to avoid the problem of applying insecticides on the
     island. However, to follow the development of the natural medfly
     population, about 10% of the sour oranges were left on selected trees.
II. The second period was devoted to repressing the reproductive rate of
     emerged flies by using massive releases of sterile insects. The timing
     of the application of the SIT was established by:

          The emergence of the first flies from pupae collected regularly
          during the winter and spring from infested sour oranges (key host
          in Procida);
          The first flies captured in the sticky traps which remained in the
          field until the first newly emerged wild flies were captured;
          The weather conditions.

     Releases were continued 2 months after the last record of flies emerging
     from pupae collected from sour oranges. During this time it was
     assumed that the flies could have about two generations in the field.
     These releases were carried out at the rate of about 10 000 sterile flies
     per week distributed first on the over-wintering medfly areas and
     afterwards on all Procida cultivated areas.
III. The third period was dedicated to checking the newly emerged adults
     which came from the infested fruits imported to the island and from the
     oviposition of gravid females which immigrated from the nearby areas.
     This was done to establish the beginning of the second release phase.
     It was accomplished by checking the flies emerging from infested fruits
     both collected in the field and found in fruit markets. Also, 300 sticky
     traps were placed in the field 20 days after the latest release of insects
     in order to reduce the number of immigrant males. The 20-day period
     was chosen because of the short life of sterile male flies during the
     summer months. All traps were removed a few days before the beginning
     of a new distribution of sterile medflies.
IV. The fourth period was devoted to repressing the reproductive rate of
     the resurging local medfly population by releasing sterile insects for
     about 2 months.
V.   The fifth period will again be dedicated to reducing the over-wintering
     medfly population.

RESULTS AND DISCUSSION

Sampling of the medfly population on Procida before applying the SIT
revealed a 4.1% infestation in sour oranges and few wild males in the sticky
traps in March. However, adverse weather conditions in spring delayed
the beginning of the sterile adult release until May 3rd, i.e. about 10 days
after the emergence of the first medfly adults from sour oranges. During
the second period of the experiment, the wild population was completely
eradicated on Procida, but since the end of August, like in 1972, a very low

TABLE II.   RESULTS OF FRUIT INSPECTION FOR MEDFLY
INFESTATION IN THE RELEASE AND CONTROL AREAS

MONTH	HOST FRUITS EXAMINED							
	APRICOT		PEACH		PEAR		FIG	
	A	B	A	B	A	B	A	B
	Island of Procida							
June	7975	0.0	13803	0.0	-	-	-	-
July	4890	0.0	14761	0.0	8725	0.0	160	0.0
August	110	0.0	12413	0.4	13240	0.0	15894	0.0
September	-	-	12478	8.0	3900	0.0	19580	0.0
October	-	-	2958	23.6	550	0.0	5148	1.3
	Monte di Procida[a]							
June	681	0.0	146	0.0	32	0.0	-	-
July	2589	6.0	362	0.8	573	0.7	/	/
August	262	80.1	58	32.7	739	17.6	/	/
September	-	-	25	64.0	360	33.0	/	/
October	-	-	-	-	477	50.1	/	/
	Island of Ischia[a]							
June	205	0.0	40	0.0	-	-	-	-
July	373	3.2	217	0.2	214	0.0	/	/
August	72	22.2	195	22.6	203	2.0	/	/
September	-	-	619	69.8	196	24.4	/	/
October	-	-	280	82.8	-	-	/	/
	Island of Capri							
June	-	-	-	-	-	-	-	-
July	385	1.0	1250	0.0	650	0.0	-	-
August	-	-	755	40.0	930	0.0	230	0.0
September	-	-	175	86.8	-	-	200	1.5
October	-	-	143	100.0	-	-	212	30.6

[a] Data from Dr. Fimiani, Istituto di Entomologia Agraria, Portici, Naples.
   A = No. of fruits examined
   B = percentage of infested fruits
   / = unknown data.

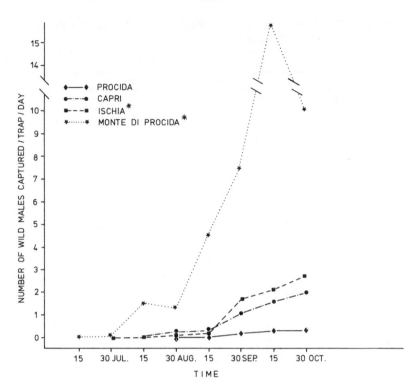

FIG. 3.   Number of captured wild males on Procida and in control areas in 1973. (Data taken from
Dr. Fimiani, Istituto di Entomologia Agraria, Portici, Naples.)

infestation was found on the island due to immigration of adults from
Monte di Procida and Ischia. In fact, the recapture of marked adults released
in the most important infested area of Monte di Procida and along the west
coast of Ischia confirmed the thesis of a continuous immigration of medfly
from these areas to Procida since August. This infestation, which was
almost exclusively limited to wild peaches, was not stopped by the second
release of sterile adults at the end of August.

However, periodic examination of host fruits in the control areas, where
medfly is controlled by insecticides, indicated that the attack on Procida was
very low (Table II). Trapping of wild adults in experimental and control
areas (Fig. 3) also showed a good result for the 1973 trial programme,
although a general delay was observed in the natural population increase
which did not reach the level of preceding years. The degree of fruit
infestation, however, confirms the great ability of this species to infest its
host even at a low population density [8]. This behaviour of the flies must
be carefully considered in planning the quantity of sterile insects to release.
This quantity should mainly assure a high probability of matings between
sterile and wild insects following good dispersal of a sufficient number of
sterile flies in the selected experimental area. In the Procida environment
this number was established to be about 10 000 adults per ha. The results

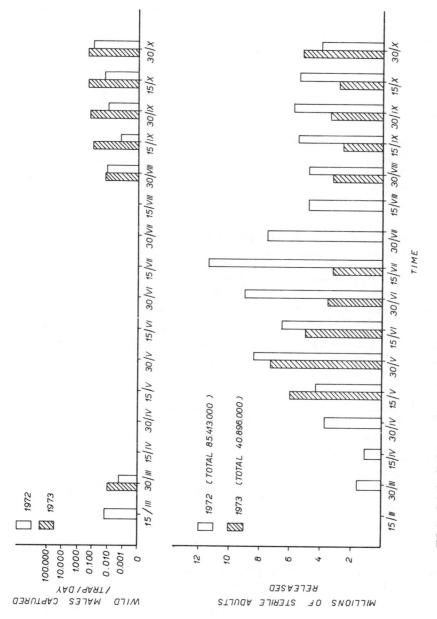

FIG. 4.   Relationship between the number of wild males captured and the numbers of sterile medfly released during the pilot experiment on Procida.

TABLE III.   DATA OF MEDFLY WILD MALES CAPTURED IN PROCIDA
ISLAND USING PLASTIC TRAPS

YEAR	CAUGHT WITH TRAP/DAY						
	JUN.	JUL.	AUG.	SEP.	OCT.	NOV.	DEC.
1968	0.25	6.23	26.66	48.63	14.59	1.56	0.21
1969[a]	0.75	1.84	2.83	51.00	25.22	1.44	0.02
1970	0.38	-	22.17	-	19.72	-	-
1971[b]	-	19.23[c]	-	68.76[c]	55.18[c]	8.64[c]	2.08[c]
1972[a]	0.00	0.00	0.01	0.07	0.17	1.87[c]	0.13[c]
1973[a]	0.00	0.00	0.01	0.15	0.33		

[a] Years of SIT application
[b] Data from Dr. de Murtas (CNEN)
[c] Capture with sticky traps.

TABLE IV.   INCIDENCE OF MEDFLY LARVAE IN PROCIDA ISLAND

YEAR	JUN.	JUL.	AUG.	SEP.	OCT.	NOV.[b]	DEC.[b]
1967	c	c	c	xxxx	xxxx	xxxx	xxxx
1968	x	xxx	xxxx	xxxx	xxxx	xxxx	xxxx
1969[a]	x	x	x	xxx	xxxx	xxxx	xxxx
1970	x	xxx	xxxx	xxxx	xxxx	xxxx	xxxx
1971	x	xxx	xxxx	xxxx	xxxx	xxxx	xxxx
1972[a]	0	0	x	x	x	x	x
1973[a]	0	0	x	x	x	x	c

a = year of SIT application;
b = attack related to the orange fertile stings;
c = fruits not inspected.
   Fruits examined: June and July — apricots and peaches; August — peaches, pears and figs;
   September and October — figs and peaches; November and December — oranges and tangerines.
   Level of infestation: 0 = zero; x = slight (< 10%); xx = moderate (10 - 30%); xxx = strong (30 - 60%);
   xxxx = very strong (> 60%).

obtained in 1973 on Procida showed that by using a strategy of releasing
considerably fewer sterile insects than in 1972, medfly control was still
very effective (Fig.4).   This was possible only because of good integration
of other control techniques and very accurate sampling of the medfly
population.   The most important advantages of this new medfly control
strategy were the reduction of the cost of the experiment and better collaboration with the farmers in trapping and sour orange destruction.   However,

some problems did arise during the 1973 application of the SIT. First, the new release procedure took more time than when the paper bags were used because few flies left the cloth cages when the car was moving. Second, it was observed that about 35% of the sterile medflies were weak and unable to fly from the cages. This clearly indicated the necessity to modify the procedure of handling the flies in the cloth cages which, in contrast, were very practical for the sterilization of large numbers of adults in a short time. Ways of improving the techniques used can be easily foreseen.

At the end of the sterile medfly release experiments on the island of Procida there were no doubts about the effectiveness of the SIT to control this species, as is shown in Tables III and IV. However, research on Procida will continue for some years to estimate the changes induced in this environment after a lengthy and strong reduction of the medfly population. Efforts will be made to have a final analysis of all results and to estimate the cost of medfly control. This information should prove very important when implementing the large new SIT programmes which CNEN recently promoted in Sardinia and Latium.

REFERENCES

[1]   PICECE, P., Report on problems dealing with the presence of Ceratitis capitata Wied. in Italy, Joint EPPO/OILB Conference on Ceratitis capitata, Madrid, 19-21 May 1970.

[2]   De MURTAS, I.D., CIRIO, U., GUERRIERI, G., ENKERLIN-S, D., "An experiment to control the Mediterranean fruit fly on the island of Procida by sterile-insect technique", Sterile-Male Technique for Control of Fruit Flies (Proc. Panel Vienna, 1969), IAEA, Vienna (1970) 59.

[3]   NADEL, D.J., GUERRIERI, G., "Experiments on Mediterranean fruit fly control with the sterile-male technique", Sterile-Male Technique for Eradication or Control of Harmful Insects (Proc. Panel Vienna, 1968), IAEA, Vienna (1969) 97.

[4]   De MURTAS, I., CIRIO, U., L'allevamento massivo della mosca della frutta (Ceratitis capitata Wied.) nell'insettario della Casaccia ed aspetti economici di lotta autocida, IX Congresso Italiano di Entomologia, Siena, giugno 1972 (in press).

[5]   CIRIO, U., De MURTAS, I., "Status of Mediterranean fruit fly control by the sterile-male technique on the island of Procida", The Sterile-Insect Technique and its Field Applications (Proc. Panel Vienna, 1972), IAEA, Vienna (1974) 5.

[6]   CIRIO, U., Influence de l'alimentation larvaire sur la longévité et la fécondité de Ceratitis capitata Wied., Joint EPPO/OILB Conference on Ceratitis capitata, Madrid, 19-21 May 1970.

[7]   CIRIO, U., De MURTAS, I., Distribuzione e ricattura di adulti radiosterilizzati di Ceratitis capitata Wied., IX Congresso Italiano di Entom., Siena, giugno 1972 (in press).

[8]   CIRIO, U., De MURTAS, I., GOZZO, S., ENKERLIN-S, D., Preliminary ecological observation of Ceratitis capitata Wied. on Procida island with a view to an attempt to control by the sterile-male technique, Boll. Lab. Entom. Agr. "Filippo Silvestri" (in press).

[9]   De MURTAS, I., CIRIO, U., ENKERLIN-S, D., Dispersion de Ceratitis capitata Wied. dans l'île de Procida (Italy), OEPP Publication, Bull. No. 6 (1972) 69.

[10] De MURTAS, I., CIRIO, U., ENKERLIN-S, D., Distribution of sterilized Mediterranean fruit fly on Procida and its relation to doses and feeding, OEPP Publication, Bull. No. 6 (1972).

[11] FIMIANI, P., Osservazioni bioecologiche sulla mosca della frutta (Ceratitis capitata Wied.) effettuate te nella zona di Monte di Procida (Napoli) negli anni 1968-1969, Boll. Lab. Entom. Agr. "Filippo Silvestri", Portici, Vol. XXX (1972) 71.

[12] FIMIANI, P., TRAFAGLIA, A., Influenza delle condizioni climatiche sull'attività moltiplicativa della Ceratitis capitata Wied., Ann. Fac. Sci. Agr. Portici (1972).

[13] FIMIANI, P., PANDOLFO, F., Fluttuazioni delle popolazioni di adulti di mosca della frutta nel litorale flegreo, Inf. tore Fitopat. 6 (1973) 13.

[14] CIRIO, U., Basi ecologiche per un programma di lotta contro la Ceratitis capitata Wied. nell'isola di Procida (in press).

GENETIC CONTROL OF Ceratitis capitata

Practical application of
the sterile-insect technique*

L. MELLADO, P. ROS, M. ARROYO, E. CASTILLO
Instituto Nacional de Investigaciones Agronómicas,
Madrid, Spain

Abstract

GENETIC CONTROL OF Ceratitis capitata: PRACTICAL APPLICATION OF THE STERILE-INSECT TECHNIQUE.
Since 1965 the INIA has been carrying out a programme of biological control of Ceratitis capitata
(Wiedemann) by means of the sterile-insect technique (SIT). Preliminary field experiments in 1969 in the
province of Murcia, Spain, showed the method to be effective when applied in small isolated areas. New field
experiments in 1972 in the province of Granada showed that it was possible to protect a semi-isolated area of
100 ha by creating peripheral "barriers" of sterile insects. No "sterile punctures" were observed in the fruit
of the release area. In 1973, mass-rearing techniques were improved and methods for shipment of refrigerated
adults instead of pupae were developed. Results of these experiments confirm (a) that the SIT is fully effective
in the control of C. capitata, (b) that shipment of insects in the adult stage is more effective than in the pupal
stage, and (c) that improvement in the rearing system can reduce the cost of production of irradiated insects
by 60%.

INTRODUCTION

1. Background

In 1965 the National Institute for Agricultural Research (INIA) started
a programme of biological control of Ceratitis capitata (Wiedemann) by the
sterile-insect technique (SIT). Its ultimate aim was to develop a technique
which could be used in suitable combination with conventional systems to
eradicate or control the pest, avoiding at the same time a series of undesirable effects (environmental pollution and disturbance of the biological
equilibrium) resulting from the exclusive use of chemical pesticides.

Between 1966 and 1969 a series of basic investigations was carried out
in order to improve the methods of large-scale rearing and sterilization
[3, 5, 7, 9]. At the same time, small-scale field applications were performed in order to study the methods of release of insects and the efficiency
of the system [1, 4, 6, 7, 9]. In 1969, in the province of Murcia, an experiment was conducted for the first time in Spain aimed at controlling the pest
in a small regular plantation of citrus and stone-fruit trees by the SIT
exclusively. The results were fully satisfactory and showed that the method
was fully effective when applied in small isolated areas [8, 9, 12]. During
1970, 1971 and 1972, laboratory investigations [11, 12] and small-scale
field experiments [2, 12] were continued, during which a study was made
of some problems which needed to be solved so that the method could be

* This work was supported by the International Atomic Energy Agency (Research Contract No. 848/R3/RB).

applied economically on a large scale, namely irradiation of adults, survival and dispersion of sterile insects in the field, methods of release, occurrence of sterile punctures in stone-fruit trees, and ecological aspects.

In 1972 a new field experiment was carried out in the province of Granada over an area of some 100 ha, where the pest was controlled entirely effectively by creating peripheral 'barriers' of sterile insects [13]. The problem of 'sterile punctures' in the fruit of the release area was eliminated fully by this procedure.

2. Present status

The present status, as resulting from the activities described, can be summarized as follows:

(a) The 1969 experiments demonstrated that in sufficiently isolated areas the application of the SIT is by itself fully effective for the control of C. capitata, even when the native population is large. The 1972 [13] and 1973 field experiments, described in the present paper, have confirmed the efficiency of the technique;

(b) A problem encountered during the 1969 experiment was the appearance of sterile punctures in stone-fruits produced by irradiated females. These punctures do not affect the quality or health of the fruit but can, to a certain extent, impair its appearance. The experiments carried out in 1970, 1971 and 1972 showed that, except in some varieties of peach, the sterile punctures were not visible upon normal inspection and did not therefore constitute a serious problem as had seemed to be the case at the beginning [10];

(c) The exclusive application of this technique as the sole method of control is, at present, excessively costly. Improvement of techniques and mechanization of operations for a large-scale programme would of course lead to a substantial reduction in costs. The use of the technique as part of a programme of integrated control, in which it would act on low-density populations, may already be economic under present circumstances.

1973 PROGRAMME

1. Objectives

(a) To demonstrate once more the efficiency of the SIT for the control of C. capitata in the area treated in 1972, with the use of improved techniques;

(b) To continue the study of the time-table of the appearance of C. capitata adults in an extensive area (some 200 000 ha) and compare it with that of the previous years for the purpose of extending the experimental area in the future;

(c) To release sterile insects at the over-wintering centres of C. capitata, especially on the coast;

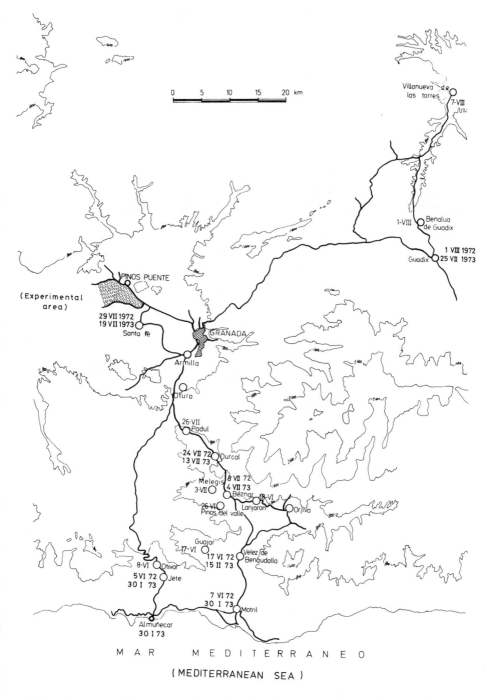

FIG.1.   General map of the area of operation and location of the experimental area.   The dates indicate
the first appearance of C. capitata adults in each of the localities in 1972 and 1973.

54 MELLADO et al.

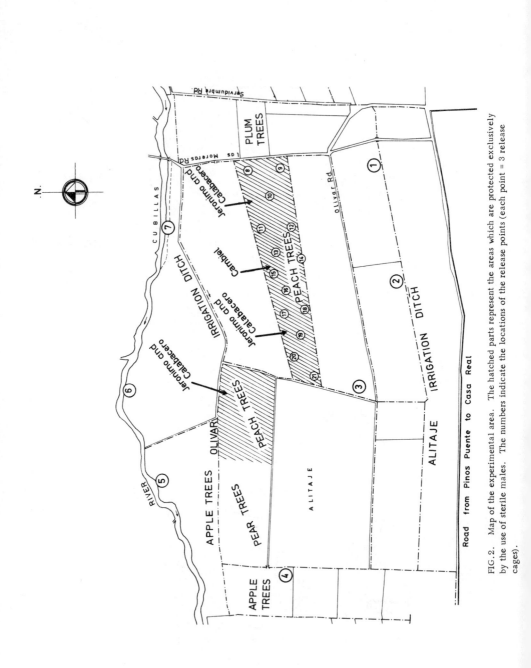

FIG.2. Map of the experimental area. The hatched parts represent the areas which are protected exclusively by the use of sterile males. The numbers indicate the locations of the release points (each point = 3 release cages).

(d) To try methods of shipment and release of refrigerated males,
     comparing the efficiency with that of shipment of pupae;
(e) To perform preliminary experiments on aerial release techniques;
(f) To improve the methods of rearing and handling of insects with
     a view to reducing costs.

2. Description of the experimental area

All field experiments were carried out in the same experimental area
as in 1972 (Fig.1), described in detail in an earlier publication [13].

In the coastal areas of the province of Granada, C. capitata is an
endemic pest, and under the climatic conditions prevailing there, the insect
exists in the adult form throughout the year.

In the interior of the province, the host plants are essentially stone-
fruit trees in regular plantations. The early-ripening fruits usually escape
infestation by C. capitata, which occurs as the season advances, presumably
as a result of the migration of insects from the warmer coastal areas [13].

The experimental area for the application of the SIT is situated at
Pinos Puente in the valley of Granada (850 m above sea level). It covers
some 100 ha, including a regular 25 ha peach plantation with some
10 000 trees in all (Fig.2). This area had already been used for experiments
in previous years. In 1971, a study was made of the cultivation and plant
health practices and data were obtained on the yield and usual losses due to
the pest. The dispersion and longevity of sterile insects were also determined
by experimental releases [12]. In 1972, this area could be protected fully
by the exclusive use of sterile insects [13].

The regular peach plantation includes the varieties Jerónimo, Calabacero
and Campiel. Campiel (600 trees) is a very late-ripening variety and is
more vulnerable to infestation by C. capitata.

The average total harvest of the above-mentioned varieties is 220 000 kg,
of which 25 000 kg is accounted for by Campiel. In 1973, owing to climatic
factors (frost), the harvest was very low — 31 000 kg, including 5500 kg of
Campiel.

The plantation is very favourably situated for experiments on biological
control by the SIT. Though lacking in topographic isolation, it is surrounded
by herbaceous crops; there are no known hosts within a radius of about 5 km,
except a few small isolated population foci.

The areas selected as control were three plantations at Armilla,
Santa Fe and Purchil. Their altitudes and climatic, topographic and
ecological conditions are entirely similar to those of the experimental area.

In a normal year, in spite of chemical treatment, C. capitata infestation
affects at least 5% of the total early output, reaching as much as 17% in the
case of the Campiel variety.

The main characteristics and treatments performed in 1973 of the
experimental and control areas are as follows:

(1) Pinos Puente (experimental area)

    Area of the peach plantation: 25 ha
    Age of trees: 6 and 9 years
    Varieties: Jerónimo, Calabacero and Campiel.

(2) Armilla (control area)

Area of the peach plantation:	10 ha
Age of trees:	6 years
Varieties:	Jerónimo and Calabacero
Treatments:	Two treatments with 0.2% Lebaycid by full spraying on 26 July and 10 August.

(3) Santa Fe (control area)

Area of the peach plantation:	8 ha
Age of trees:	7 years
Varieties:	Jerónimo and Calabacero
Treatments:	Two treatments, one on 1 August with 0.2% Lebaycid by full spraying and the other on 11 August with Lebaycid + 0.5% buminal in strips.

(4) Purchil (control area)

Area of the peach plantation:	9 ha
Age of trees:	6 years
Varieties:	Jerónimo, Calabacero and Campiel
Treatments:	Two treatments with 0.2% Lebaycid by full spraying on 26 July and on 10 August for early-ripening varieties. Two treatments with 0.2% Lebaycid by full spraying on 29 August and 13 September for the late Campiel variety.

3. Material and methods

3.1. Capture of native insects

Plastic fly-traps containing an attractant (Trimedlure) and an insecticide (DDVP) were set up at the localities Jete, Motril, Almuñecar, Velez de Benaudalla, Beznar, Durcal, Santa Fe and Guadix (Fig. 1).

Four traps were set up at each of the above localities in regular peach plantations. (In some of them other fruit trees are also present, namely orange, sapodilla and apricot.) All traps were inspected twice a week and the inspection was continued until the presence of the insect was confirmed.

3.2. Release of sterile insects

The release of sterile insects began towards the end of March at Almuñecar, Jete, Motril and Velez, since (in accordance with the time-table of appearance of Ceratitis established in 1972 [13]) these places were regarded as over-wintering centres of the fly. From 11 July onwards all releases were carried out in the experimental area of Pinos Puente.

All the insects came from the mass artificial rearing laboratory in Madrid (El Encín) and were supplied in the pupal or in the adult stage. Releases in the pupal form were carried out by the method used in 1972 [13]. The adults were released at suitable places in the field where the sun could rapidly revive them after they had been kept at a low temperature for several hours.

All releases in the experimental area were carried out at points 1 to 7 (Fig. 2), except in the case of the late-ripening variety Campiel, for which the insects were released in the interior of the plot. The time-table of releases and the number of insects released are given in Table I.

3.3. New method of shipment of pupae

C. capitata pupae were sent from the mass-rearing centre to the experimental area in cages made of a parallelepiped wooden frame with very fine wire netting on all sides.

Boxes with pupae (100 000) are stacked in these containers. The top of the boxes consists of plastic netting through which adults emerging during the journey can leave the boxes and disperse through the whole volume of the container. In this manner, when the shipment arrives in the field, about 45% of the insects are in the adult stage and in full activity.

3.4. Laboratory techniques for obtaining irradiated adults

In order to store the largest number of irradiated adults in the smallest possible space, we made some cylinders of cloth netting, 2.50 m in height and 45 cm in diameter. The lower part contains a number of circular plastic trays and the upper part a funnel, the major diameter of which is equal to that of the lower tray.

The bottom of the tray contains a layer of sawdust and above that a layer of pupae (50 000). The whole tray is covered with a circular lid of wire netting, the openings in which allow the newly-emerged adults to pass but hold back the pupae.

These cylinders are suspended from the roof. As adults emerge from pupae, they pass through the netting, fly upwards and scatter over the whole surface of the cylinder.

When there is a sufficient quantity of adults, the cages are folded up, placed in a refrigerated chamber and kept there for some time until the adults lose activity. Then the cages are again suspended from the roof, but now in the inverted position. The adults slip down along the fabric and fall into the funnel, which directs them into appropriate containers. The pupae cannot fall since they are retained by the wire netting covering the tray.

The system described is simple, inexpensive and easy to operate. One can use either previously irradiated pupae or non-sterile pupae if it is desired to irradiate them in the adult stage.

In the former case (method used for releases in 1973), the adults were collected in cylindrical containers with very fine metal gauze bases having a capacity of 30 000 insects.

TABLE I. RELEASE OF STERILE INSECTS

Date	Number		% emergence from pupae	% mortality of adults	True equivalence of useful adults released	Place of release
	Pupae	Adults				
March 29	1 000 000	600 000	58	4	1 156 000	Jete
April 3	1 000 000	400 000	60	2	992 000	Velez
April 26	1 000 000	500 000	48	3	965 000	Jete
May 3	1 000 000		51		510 000	Motril
May 8	1 000 000	300 000	45	5	735 000	Almuñecar
May 17		600 000		8	552 000	Jete
May 29		1 000 000		12	880 000	Motril
June 12		250 000		8	230 000	Velez
June 14		400 000		50	200 000	Jete
June 15	1 000 000	600 000	67	10	1 210 000	Jete
June 16	1 000 000	800 000	70	20	1 340 000	Velez
June 20	700 000	350 000	68	25	738 500	Jete
June 23	750 000		65		487 500	Velez
June 27	700 000	120 000	68	40	548 000	Velez
June 28		450 000		10	405 000	Melegís
June 29		450 000		40	270 000	Melegís
July 3	750 000	300 000	68	3	801 000	Pinos del V.
July 4		600 000	50		300 000	Beznar
July 6	800 000		55		440 000	Melegís
July 7	1 000 000	200 000	61	8	794 000	Beznar
July 11		500 000		20	400 000	Experimental plot
July 12		450 000		40	270 000	"
July 13	800 000	200 000	58	15	634 000	"

Date	Number		% emergence from pupae	% mortality of adults	True equivalence of useful adults released	Place of release
	Pupae	Adults				
July 14	800 000	200 000	56	10	628 000	Experimental plot
17		500 000		10	450 000	"
20	750 000		55		412 500	"
21	1 000 000	450 000	53	20	890 000	"
24	800 000	300 000	65	10	790 000	"
28	1 000 000		55		550 000	"
31	1 000 000	500 000	61	8	1 070 000	"
August 1	1 100 000	700 000	68	10	630 000	"
2	1 000 000	900 000	62	10	1 558 000	"
4	1 000 000	150 000	68	8	758 000	"
7		500 000		40	980 000	"
8		200 000		8	184 000	"
9		550 000		10	500 000	"
10		300 000		10	270 000	"
11	800 000		64		512 000	"
14	1 250 000		52		650 000	"
17	2 500 000	1 000 000	68	35	2 350 000	"
18	1 500 000		65		975 000	"
21	2 400 000	1 500 000	55	50	2 070 000	"
23	1 100 000	500 000	50	10	1 000 000	"
24	1 100 000		62		680 000	"
25	900 000		55		495 000	"
28	1 000 000		60		600 000	"
29	2 000 000	600 000	68	8	1 912 000	"
30		800 000		10	720 000	"
31	1 000 000		68		680 000	"

TABLE I. RELEASE OF STERILE INSECTS (Cont.)

Date	Number		% emergence from pupae	% mortality of adults	True equivalence of useful adults released	Place of release
	Pupae	Adults				
September 1	2 700 000		50		1 350 000	Experimental plot
4	1 650 000	300 000	60	8	1 266 000	"
6	1 000 000	800 000	55	10	1 270 000	"
8	2 200 000		62		1 364 000	"
11	1 100 000	600 000	65	15	1 225 000	"
12	1 000 000	200 000	62	4	812 000	"
13		600 000		12	528 000	"
15	600 000		59		354 000	"
16	600 000		60		360 000	"
19	800 000	200 000	68	5	734 000	"
20	1 000 000		60		600 000	"
22	1 500 000		65		975 000	"
25		600 000		60	240 000	"
26	1 100 000	200 000	68	35	878 000	"
28	2 350 000		50		1 175 000	"
October 3		800 000		60	320 000	"
TOTAL	54 100 000	23 020 000			50 625 500	

3.5. Techniques of shipment of refrigerated adults

After the containers have been filled with adults, they are placed in portable ice-boxes with several bags of ice. Thus packed they are sent by night train to the experimental area. On arrival, the adults are released at suitable points in the sun so that they rapidly regain their body temperature and can resume their normal activity in the shortest possible time.

In practice, this system of shipment has been found to be very efficient and, save for a few accidents (bursting of ice bags), the percentage of mortality has always remained within very low limits (see Table I).

Depending on the ambient temperature during the journey to the experimental area, a greater or smaller number of ice bags are used so that the temperature inside the box remains between 5 and 8°C.

3.6. Preliminary experiments on aerial releases

In the present experiment we carried out several trials of aerial release from Piper aircraft. The primary aim is to find a simple, economical method which could be applied on a large scale by non-specialized personnel.

During these trials the support material for Ceratitis adults was cork shavings and small polystyrene balls. The latter material was found to behave better. Altogether 2 500 000 adults were used in these experiments.

The results are promising but the system requires improvement, since very high mortality was observed.

3.7. Inspection of fruits

In the experimental and control areas, samples of fruits were taken in the process of harvesting (including fruits which had fallen to the ground). Samples were collected from crates at random on different days of harvesting. The details of the number of fruits inspected in each sampling operation and of the total harvest in each case are given in Tables II and III.

In the case of the late-ripening variety Campiel, samples were taken from the fruit-receiving bench and as the fruits were being packed in crates (Table IV).

4. Results

Figure 1 shows the dates of appearance of C. capitata adults in each of the localities mentioned.

Tables II and III give data on C. capitata infestation of earlier ripening peaches in the control (Armilla, Santa Fe and Purchil) and in the experimental (Pinos Puente) areas.

Table IV presents data on C. capitata infestation of the late-ripening peach variety Campiel in the control (Purchil) and in the experimental (Pinos Puente) areas.

These tables show that:

1. In all samples, infestation by C. capitata of the earlier varieties of peach (Jerónimo and Calabacero) was less than 1.3% in the experimental area, which was protected exclusively by means of sterile-insect release.

In samples taken from the control areas (treated chemically) infestation
varied between 1.2% and 17%. In non-treated areas, it lay between
92% and 100%.

2.   The following averages were obtained for infestation:
Early-ripening varieties

        Experimental area   0.12%

        Control areas:

        Santa Fe   2.20%
        Purchil   8.30%
        Armilla   6.70%

Late-ripening varieties

        Experimental area   0.30%
        Control area (Purchil)   2.10%

3.   It can be concluded from Table II that shipment of insects in the adult
stage (refrigerated) is more effective than in the pupal stage. These
results can be summarized in the following figures (the 1972 data are
cited for comparison):

	1972	1973
Total shipment of pupae (millions)	40.7	54.1
Total useful insects (millions)	25.0	32.4
Efficiency of the method (%)	61.3	60.0
Total shipment of adults (millions)		23.0
Total useful insects (millions)		18.1
Efficiency of the method (%)		79.0

The above data indicate that shipments of pupae give about 60% useful
insects, as compared with 79% in the case of shipments of adults.

TABLE II. PERCENTAGES OF Ceratitis capitata INFESTATION IN THE
EXPERIMENTAL AREA
(Jerónimo and Calabacero varieties)

Place	Date	Fruit harvested (kg)	Fruit infested (kg)	%
Pinos Puente	4, 5, 6/8	5 000	0.5	0.01
	7/8	1 500	20	1.30
	8/8	5 000	0.5	1.01
	14/8	4 000	1	0.02
	15/8	3 500	1.5	0.04
	16/8	2 000	1	0.05
	17/8	2 000	2	0.10
	21/8	1 000	–	0.00
	29/8	1 500	5	0.30

TABLE III. PERCENTAGE OF Ceratitis capitata INFESTATION IN
CONTROL PLOTS
(Jerónimo and Calabacero varieties)

Place	Date	Fruit harvested (kg)	Fruit infested (kg)	%	Remarks
Santa Fe	30/7	1 100	28	2.50	2 treatments
	31/7	360	6	1.70	(Lebaycid) on
	7/8	3 000	49	1.60	1 Aug. and 11 Aug.
	8/8	1 500	18	1.20	
	9/8	500	30	6.00	
	10/8	1 700	46	2.70	
	11/8	2 000	39	1.85	
	13/8	1 700	41	2.40	
	15/8	500	20	4.00	
Purchil	3/8	1 250	44	3.50	2 treatments
	4/8	2 480	97	3.90	(Lebaycid) on
	5/8	2 230	101	4.50	26 Jul. and 10 Aug.
	6/8	3 450	210	6.10	
	7/8	1 860	54	2.90	
	8/8	2 660	98	3.70	
	15/8	3 845	357	9.30	
	16/8	2 300	200	8.70	
	17/8	3 640	375	10.30	
	20/8	5 435	717	13.20	
	24/8	1 824	312	17.10	
Armilla	31/7 - 5/8	12 600	542	4.30	2 treatments
	7/8 - 10/8	6 348	387	6.10	(Lebaycid) on
	11/8 - 17/8	8 300	482	5.80	26 Jul. and 10 Aug.
	17/8 - 24/8	6 840	875	12.80	
Purchil	10/8	3 224	2 960	92.00	No chemical treatment
Armilla	6/8	2 840	2 840	100.00	No chemical treatment

4. Improvement in the rearing systems has reduced the cost of production
of irradiated insects in 1973 (in comparison with that in 1972) by
approximately 60%.
5. It has not yet been possible to develop a satisfactory method of serial
release.

TABLE IV. PERCENTAGE OF Ceratitis capitata INFESTATION
(Late-ripening variety Campiel)

Place	Date	Number of fruits harvested	Number of fruits infested	%	Remarks
Purchil	26/9	8 400	53	0.63	Control area.
	26/9	300	18	6.00	Treatment on
	27/9	5 280	110	2.08	29 Aug. and 13 Sep.
	2/10	7 580	93	1.22	
	3/10	10 370	122	1.17	
	4/10	7 160	83	1.18	
	7/10	7 020	89	1.26	
	8/10	7 100	98	1.38	
	14/10	2 820	219	7.70	
	15/10	2 960	339	11.50	
Pinos Puente	25/9	7 954	4	0.05	Experimental area.
	26/9	7 250	21	0.28	No treatment.
	27/9	1 920	4	0.20	
	30/9	4 520	18	0.39	
	2/10	7 680	21	0.27	
	3/10	2 340	25	1.06	
	7/10	7 548	28	0.37	

CONCLUSIONS

The results obtained in 1973 confirm those of 1972 and 1969 in the sense
that the sterile-insect technique is fully effective against C. capitata. The
reduction in the cost of production and the higher efficiency of the method
of release of refrigerated adults represent a substantial progress in the
practical application of the technique on a large scale.

REFERENCES

[1] MELLADO, L., CABALLERO, F., ARROYO, M., JIMENEZ, A., "Ensayos sobre erradicación de
    Ceratitis capitata Wied. por el método de los 'machos estériles' en la isla de Tenerife" (Experiments
    on the eradication of C. capitata Wied. by the sterile-male technique on the island of Tenerife), Boletín
    de la Estación de Fitopatología Agrícola, no.399, INIA, Madrid (1966).
[2] ARROYO, M., MELLADO, L., JIMENEZ, A., ROS, P., "Lucha biológica contra Ceratitis capitata por
    el método de 'machos estériles'" (Biological control of C. Capitata by the sterile-male technique),
    Progress Report 1971, INIA, Madrid.
[3] ARROYO, M., JIMENEZ, A., MELLADO, L., CABALLERO, F., "Aplicación de isótopos radiactivos
    a la investigación de métodos sobre lucha biológica contra plagas (Application of radioisotopes in the
    study of methods of biological control of pests),
    I. Ensayos sobre marcado de adultos de Ceratitis capitata con P-32 (Experiments on ^{32}P-labelling of
        C. capitata adults).
    II. Ensayos sobre marcado de larvas de Ceratitis capitata con P-32 (Experiments on ^{32}P-labelling of
        C. capitata larvae).

III. Obtención de machos estériles de Ceratitis capitata mediante la irradiación de sus pupas con rayos gamma (Production of C. capitata sterile males by gamma irradiation of pupae).

IV. Efectos de las radiaciones gamma sobre pupas de Ceratitis capitata en función del fraccionamiento de la dosis de irradiación (Effects of gamma radiation on C. capitata pupae as a function of irradiation dose fractionation).

V. Efectos de la radiación gamma sobre pupas de Ceratitis capitata previamente marcadas con P-32 " (Effects of gamma radiation on C. capitata pupae labelled in advance with ^{32}P),

Boletín de la Estación de Fitopatología Agrícola, Vol.XXVIII, no.388-392, INIA, Madrid (1965).

[4] MELLADO, L., ARROYO, M., JIMENEZ, A., CABALLERO, F., "Ensayos sobre erradicación de Ceratitis capitata Wied. por el método de los 'machos estériles' en la isla de Tenerife. Progresos realizados durante el año 1967" (Experiments on the eradication of C. capitata Wied. by the sterile-male technique on the island of Tenerife. Progress report 1967), Boletín de la Estación de Fitopatología Agrícola, Vol.XXX, INIA, Madrid (1967-68).

[5] ARROYO, M., MELLADO, L., JIMENEZ, A., CABALLERO, F., "Influencia en la puesta de Ceratitis capitata Wied. de distintas dietas alimenticias" (Influence of different diets on the oviposition of C. capitata Wied.), Boletín de la Estación de Fitopatología Agrícola, Vol.XXX, INIA, Madrid (1967-68).

[6] MELLADO, L., ARROYO, M., JIMENEZ, A., CABALLERO, F., "Ensayos sobre erradicación de Ceratitis capitata Wied. por el método de 'machos estériles' en la isla de Tenerife. Progresos realizados en 1968" (Experiments on the eradication of C. capitata Wied. by the sterile-male technique on the island of Tenerife. Progress report 1968), Boletín de Patología Vegetal y Entomología Agrícola, Vol.XXXI, INIA, Madrid (1969).

[7] MELLADO, L., ARROYO, M., JIMENEZ, A., "Work at the Instituto Nacional de Investigaciones Agronómicas", Sterile-Male Technique for Control of Fruit Flies (Proc. Panel Vienna, 1969), IAEA, Vienna (1970) p.90.

[8] MELLADO, L., NADEL, D.J., ARROYO, M., JIMENEZ, A., "Mediterranean fruit fly suppression experiment on the Spanish mainland in 1969", Sterile-Male Technique for Control of Fruit Flies (Proc. Panel Vienna, 1969), IAEA, Vienna (1970) 91.

[9] MELLADO, L., "La técnica de machos estériles en el control de la mosca del Mediterráneo: programas realizados en España (The sterile-male technique in the control of the Mediterranean fly: survey paper), Sterility Principle for Insect Control or Eradication (Proc. Symp. Athens, 1970), IAEA, Vienna (1971) 49.

[10] MELLADO, L., ARROYO, M., JIMENEZ, A., CASTILLO, E., "Ensayos de lucha autocida contra Ceratitis capitata Wied. Progresos realizados en 1969" (Experiments on autocide control of C. capitata Wied. Progress report 1969), Anales Inst. Nac. Inv. Agrarias, Serie Protección Vegetal no.2, INIA, Madrid (1972).

[11] VARGAS, C., MELLADO, L., "Efectos del tratamiento combinado de refrigeración e irradiación gamma en adultos de Ceratitis capitata Wied" (Effects of combined treatment with refrigeration and gamma irradiation on Ceratitis capitata Wied. adults), Anales Inst. Nac. Inv. Agrarias, Serie Protección Vegtal no.2, INIA, Madrid (1972).

[12] ARROYO, M., MELLADO, L., JIMENEZ, A., CASTILLO, E., "Ensayos de lucha autocida contra Ceratitis capitata Wied. Progresos realizados en 1970" (Experiments on autocide control of Ceratitis capitata Wied. Progress report 1970), Anales Inst. Nac. Inv. Agrarias, Serie Protección Vegetal no.2, INIA, Madrid (1972).

[13] MELLADO, L., ARROYO, M., ROS, P., "Control of Ceratitis capitata Wied. by the sterile-male technique in Spain", The Sterile-Insect Technique and its Field Applications (Proc. Panel Vienna, 1972), IAEA, Vienna (1974) 63.

APPLICATION OF THE STERILE-INSECT TECHNIQUE FOR CONTROL OF MEDITERRANEAN FRUIT FLIES IN ISRAEL UNDER FIELD CONDITIONS

S.S. KAMBUROV, A. YAWETZ
Biological Control Institute,
Citrus Marketing Board of Israel,
Rehovot, Israel

D.J. NADEL
International Atomic Energy Agency,
Vienna, Austria

Abstract

APPLICATION OF THE STERILE-INSECT TECHNIQUE FOR CONTROL OF MEDITERRANEAN FRUIT FLIES IN ISRAEL UNDER FIELD CONDITIONS.
A large-scale field experiment, carried out in 1973 in Israel, employing the sterile-insect technique against the Mediterranean fruit fly and conducted over a 10 000 dunam area containing commercial citrus groves, is discussed. The release area was surrounded by a 500-m-wide low-volume (LV) bait spray barrier. Sterile flies were released from the ground and by air twice weekly. Results indicate successful control of the wild fly population for several months only and a clear suppression until July; thereafter, wild fertile females immigrated into the release area through the LV barrier.

INTRODUCTION

The Mediterranean fruit fly Ceratitis capitata (Wiedemann) is regarded as a major pest of citrus and deciduous fruit in Israel. In 1957, the Agro-technical Division of the Israel Citrus Marketing Board began an intensive study, under field conditions, of control of the wild fly population in citrus, using Malathion bait applied as a ground spray [1]. During the 1958-1959 season, experiments were carried out involving spraying SIB7 and Malathion from the air (3 litres/dunam[1]). To obtain a measure of the fly population and the efficiency of the bait spray, traps containing 'Siglure' (U.O.P. Chemicals, USA) mixed with 3% DDVP[2] were placed. In 1960-1962 'Medlure'[3] (U.O.P. Chemicals, USA) was successfully introduced as a male attractant, and the quantity of material sprayed from the air was reduced from 3 to 1 litre per dunam. During the 1965-1966 citrus season, a locally-developed bait, 'Nasiman' (Tamogan, Israel), mixed with technical-grade Malathion, was air-sprayed using the low-volume (LV) spray method. The quantity of spray, consisting of 25% Malathion and 75% bait, applied to the citrus groves was reduced to 100 ml/dunam.

---

[1] 1 dunam = 1000 m^2.

[2] DDVP = 0, 0-Dimethyl-0-2,2-dichlorvinyl-phosphate.

[3] Later replaced by 'Trimedlure'.

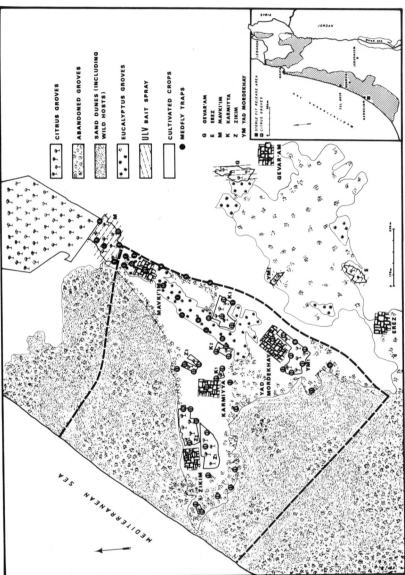

FIG.1. Sketch of the 1973 sterile-insect release experimental region.

The use of this method for Mediterranean fruit fly control in citrus by the Citrus Marketing Board (CMB) resulted in fruits almost free of infestation, achieved at a comparatively low cost and by using only small quantities of insecticide per dunam. The most important feature of the method is that it presents minimum hazard to beneficial insects which interact with various other citrus pests. It is, however, traditional in Israel to use biological control methods for most of the important citrus pests by the mass rearing and release of various beneficial insects and by spraying with selective materials. This is carried out by the Biological Control Institute of the CMB. The disadvantages of the use of LV spray in commercial citrus groves and the difficulty of controlling the wild fly population in settlements by air-distributed bait spray, forced the CMB to search for other methods of controlling the fly population, to eliminate the danger from the use of insecticides in commercial citrus groves in urban areas, and to reduce the widespread use of Malathion, even at a low dosage. The sterile-insect technique can be utilized successfully in urban areas and also in the commercial citrus and deciduous groves, reducing or eliminating insecticide pollution.

In 1972, the International Atomic Energy Agency, through the United Nations Development Programme, initiated a co-operative project with the Citrus Marketing Board and the Biological Control Institute of the Board to determine the feasibility of substituting the sterile-insect technique for the LV bait spray method for control of the Mediterranean fruit fly.

In the summer of 1972, a preliminary test of the system was made in a village about 3 600 dunams in area. Sterile males were released wherever citrus and other deciduous fruit hosts were abundant. However, severe suppression of the fly population was achieved for only about one month because of the dense wild fly population in surrounding groves and, hence, the vast immigration of fertile females into the treated area in the absence of any bait spray barrier.

THE 1973 STERILE-INSECT TECHNIQUE PROJECT

Methods and materials

The experiments were designed to demonstrate medfly suppression on a larger scale than in the trials of 1972. They were carried out in the southern coastal area of the country. The selected area of about 10 000 dunams included commercial citrus groves of all the major citrus varieties, four collective settlements and wild vegetation, including abandoned groves. The area was isolated on the north and south sides by 4 - 5 km of sand dunes and on the west by the Mediterranean sea, and was bordered on the east by cultivated fields and some commercial citrus groves (Fig. 1).

The groves outside the sterile-insect fly release area were treated every 10 - 12 days with LV bait spray to suppress migration of wild flies into the release area. Before the end of the picking season in April, grapefruit and valencia orange trees were marked in the release and bordering LV spray areas[4]; fruit on these trees was left unpicked until

---

[4] The width of the LV border was about 500 m; it was sprayed in 25-m-wide swaths at 60- to 80-m spacings.

July — August to determine the effectiveness of the sterile-insect technique
in these experiments. However, in practice, a substantial amount of fruit
was left unpicked in all citrus groves — both in the release area and outside —
because of requirements for freezing of fruits. Large quantities of these
fruits outside the release area were taken as controls.

Forty-six fly traps (Fig. 1, Traps Nos 1-46) were dispersed throughout
the release area to check periodically the distribution of the released
sterile flies, and also outside the LV spray barrier area so as to obtain
indications of wild fly population fluctuations. In addition, at the end of
July another 14 Nadel-type traps (Fig. 1, Traps A-N) were placed at various
points in a eucalyptus grove in the north part of the release area so that
information could be obtained on the migration of colour-marked (dyed) flies
released on the north border.

On the basis of the fly trap data, the release and barrier areas were
sprayed twice with LV bait spray — at the beginning of September and the
end of November, 1972. During January - March, 1973, the fly population
was low because of the low temperatures. Two weeks before the first
release of sterile males in the middle of April, the release and border
areas were sprayed by air for elimination of any fertile adult females.

Mass-rearing technique

Oviposition cages

The Seibersdorf cylindrical cloth oviposition cage [2] was replaced
by a rectangular aluminium frame cage (2 m × 2 m × 0.2 m), two sides of
which were covered with a synthetic cloth screen (16 mesh/cm), providing

FIG. 2. Rectangular oviposition cages.

FIG.3. Self-stacking larval rearing trays.

oviposition sites (Fig. 2).   From about 250 000 adults in each cage,
approximately one litre of eggs was collected every ten days.   The rearing
conditions were: temperature 25 ± 2°C; 65 ± 5% RH and continuous lighting.
The eggs were collected daily and held for 2 days in 3-litre round-bottom
flasks, aerated by aquarium stones, at a temperature of 24 ± 5°C.   The
eggs were then distributed on the artificial diet one day before hatching.

Larval medium

      For rearing the larvae, self-stacking trays of fibreglass
(0. 8 m × 0. 4 m × 0. 02 m) were used (Fig. 3), each tray holding about
100 000 larvae.   The cultures were held at 26 ± 1°C for the first 3 days
after seeding and then transferred to 20 ± 1°C and 85 ± 5% RH until fully
developed.   The mature instar larvae jumping into water were collected
every 8 hours, thus keeping the number of dead pupae to a minimum.

Pupation and handling of pupae

Mature larvae are spread in a layer about 1 cm in depth in wooden
boxes (90 cm × 60 cm × 10 cm). Pupation takes place in the dark at 25°C
and 70% RH. After 48 hours, the pupae are removed and placed on screen-
bottomed trays (90 cm × 60 cm × 2 cm), 5 litres of pupae in each, and held
at 20°C and 75% RH.

Irradiation of pupae

The pupae are irradiated by a ^{60}Co gamma irradiator (Gammacell 220)
supplied by the IAEA, at a dosage of 9 000 rad. The irradiation is performed
on the first day of emergence of the adult fly, in batch sequences every
8 hours. At 9 kR, the average male fertility was found to be about 0.7%.
Some individual males may exhibit a higher degree of fertility, but not

FIG.4. Adult emergence cabinet.

greater than 1.4%. The fail-safe instrument, connected to the Gammacell, prevented any inadvertent confusion between irradiated and non-irradiated batches of pupae.[5]

Emergence of irradiated adults, packing and release

The irradiated pupae are held in iron cabinets in screen-bottomed trays. The adults emerge in a 10 m × 6 m cloth sleeve attached to the cabinets (Fig. 4). Twice a day the flies are collected from the sleeve by chilling to 4°C, and are stored at this temperature for 2 1/2 days in netted trays until their release. One hour before release, the flies are packed into 1-gallon ventilated plastic containers. The transportation to the release area is made in insulated boxes.

The release is made from the ground or from a fixed-wing 'Cheroka Six' plane at a height of 70 - 80 m and an air speed of 70 - 75 miles/hour. The flies are released freely through a 3 1/2-inch plastic tube fixed to the floor of the aircraft. The flies are not harmed by the free fall and after 5 - 10 min commence their natural flight.

TABLE I. NUMBER OF FLIES RELEASED IN THE 1973 STERILE-INSECT TECHNIQUE TRIALS

Month	Adult		Pupae	
Number of flies released	by air	from ground	by air	from ground
April	-	6 500 000	-	-
May	10 575 000	315 000	1 210 000	1 705 000
June	13 783 000	202 000	2 750 000	-
July	20 205 000	-	16 390 000	4 400 000
TOTAL	51 580 000		26 455 000	

RESULTS AND DISCUSSION

The wild fly population generally declined in February and March. In April, when sterile fly release starts, trapping of wild flies was close to zero. All flies in April were released from the ground, while in May and June the flies were released mostly by air. Some experiments involving release of pupae by air were also carried out (Table I).

The effect of sterile flies on the wild population was measured periodically by checking samples of un-picked fruit twice a week and by collecting dropped fruit for laboratory examination.

---

[5] This instrument was constructed in the IAEA Laboratories, Seibersdorf.

KAMBUROV et al.

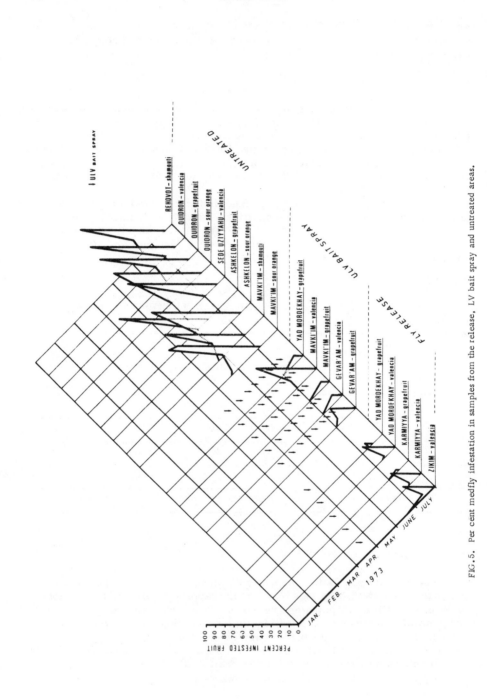

FIG. 5. Per cent medfly infestation in samples from the release, LV bait spray and untreated areas.

TABLE II. INFESTATION OF FRUIT VARIETIES IN THE BARRIER
AREA ON SPECIFIC DATES

Place	Variety	Infestation (%)	
		13 July 1973	30 July 1973
Gvar Am	Grapefruit	8.7	27.73
	Orange (valencia)	-	11.71
Mavkiym	Grapefruit	-	8.82
	Orange (valencia)	14.2	65.35
Yad Mordechai	Grapefruit	-	11.43

Beginning in May, the population of wild flies outside the LV bait spray
border increased; hundreds of flies developing on sour orange and other
citrus fruits were trapped weekly. In the first ten days of June, thousands
of infested fruits in the control area fell. In the middle of June, flies were
captured in the LV border traps, and at the end of this month about 1%
infestation was noted in samples collected from Zikim in the release area
(591 fruits).

At the same time, all samples from Karmiyya and Yad Mordechai in
the release area (valencia and grapefruit) were clean. During the first
ten days of July in Karmiyya, out of 767 grapefruit collected, 1.6% were
infested. In the middle of the same month, 8.5% infested grapefruit were
found in Yad Mordechai (out of 11 430 fruits sampled) and 5% of valencia
oranges from Zikim were infested in a sample of 4 365 fruits.

From data collected on 30 July, 1973, the damage was found to have
increased to about 34 - 39% in the whole release area (Fig. 5). On the same
day, hundreds of fruits from the LV bait spray area were collected and
checked. Higher infestation was found in the Mavkiym citrus groves, which
are nearest to the release area (Table II).

It was clear that the LV spray strip of 500 m width was not sufficient
to present a barrier to the large wild fly population, and many fertile
females penetrated into the release area. Probably most of the wild flies
infiltrated through the eucalyptus groves in the north part of the release
area where the highest infestation was found and where the distance between
the barrier and the release area was only 120 m. Since the wild fly population was increasing in the release area and the LV spray barrier had proved
inefficient, release of sterile flies was stopped at the end of July.

Experiments were made to identify the migration of wild flies to the
release area on the north boundary. Nadel traps were placed throughout the
eucalyptus groves (traps A - N, Fig. 1). Eight hundred thousand green-
coloured flies were released from the ground at the Mavuiym border line;
the marked flies were trapped after six days in large numbers, not only in
the traps in the eucalyptus grove, but also in traps placed in Zikim,
Karmiyya and Yad Mordechai and to the south sand dune border (traps
Nos 10 and 11). One week later, eight hundred thousand orange-coloured
flies were released from the ground in the same place, one day after

application of LV spray.   These flies were caught in the Nadel traps in the eucalyptus groves and also in traps Nos 10, 11, 21, 25 and 36.   These two experiments demonstrate the migration of wild flies across the entire release area from the north, in the wind direction.

The results of this experiment and those of the previous year indicate the difficulties in the use of sterile flies for controlling the wild fly popula-tion in small areas, even when release starts at the time of natural decline in the wild fly population.   Isolation of such areas with LV or ground sprays is feasible provided that wide areas surrounding the release area are treated and the wild flies are completely controlled, as in the centrally-organized fly control system in operation in Israel during the citrus season.   Under our local conditions, it is quite difficult to use the sterile-insect technique in late spring and summer because it would entail the stoppage of chemical fly control in the citrus areas.   However, good chances for successful fly control with sterile flies could be achieved in large areas which are more or less naturally isolated by releasing the irradiated flies in the autumn, when the wild fly population in citrus groves throughout the country falls under LV bait spray control.

REFERENCES

[1]   COHEN, I., COHEN, J., Centrally organized control of the Mediterranean fruit fly (Ceratitis capitata
        Wied.) in citrus groves in Israel, Agrotechnical Division, Citrus Marketing Board of Israel (1967).
[2]   NADEL, D.J., "Current mass-rearing techniques for the Mediterranean fruit fly", Sterile-Male Technique
        for Control of Fruit Flies (Proc. Panel Vienna, 1969), IAEA, Vienna (1970) 13.

APPLICATION OF SIT ON THE EUROPEAN CHERRY FRUIT FLY, Rhagoletis cerasi L., IN NORTHWEST SWITZERLAND

E.F. BOLLER, U. REMUND
Swiss Federal Research Station for
Arboriculture, Viticulture and Horticulture,
Wädenswil, Switzerland

Abstract

APPLICATION OF SIT ON THE EUROPEAN CHERRY FRUIT FLY, Rhagoletis cerasi L., IN NORTHWEST SWITZERLAND.
The procedures used for rearing, sterilizing, marking and release of Rhagoletis cerasi L. are reviewed.
A small-scale control programme was started in 1973 in two pilot orchards, to be expanded in 1974, in
order to investigate the feasibility of eradicating R. cerasi by SIT. The results achieved so far and the
problems to be solved are discussed briefly.

1. INTRODUCTION

In 1969 suitable experimental orchards were selected in the cherry
producing centre of northwestern Switzerland and ecological and behavioural
studies were carried out in anticipation of a future release programme [1,2].
Density maps were established in three potential release areas in 1969,
1970 and 1971 and a joint Swiss-Austrian collection campaign was put in
operation in 1971 in order to provide the necessary fly material for field
releases. In 1972, a first release programme involving some 600 cherry
trees and 180 000 flies was carried out in order to study the techniques
developed in the laboratory and the logistic problems connected with handling
larger numbers of flies. The experiment indicated that the initiation of an
actual control programme aiming at the eradication of the target species
was technically feasible. However, unexpected ecological factors (heavy
damage to the cherry crop due to late frosts and subsequent change of the
dispersive behaviour of the flies) pointed to the necessity of carrying out
the pilot studies in several small orchards rather than in the 50-ha area
previously selected for the control programme [3].

Two small orchards (15 and 30 trees respectively) were therefore chosen
for the control programme initiated in 1973 and a third orchard (50 trees)
was prepared for 1974 by suppressing the wild population by means of a
mass-application of visual traps [4, 5].

The objective of the investigations in progress is to prove the feasibility
of eradicating Rhagoletis cerasi on a small scale under the specific conditions
of Switzerland as a model for larger operations in the future.

2. PRODUCTION AND PREPARATION OF STERILE FLIES

2.1. Rearing

Although the current release programme is based on field-collected
material by adopting the approach suggested by the IAEA [1, 6], research

TABLE I.   COMPOSITION OF DIET G6D4 IN PER CENT

Nutrients	
Crude sugar	10.0
Brewer's yeast	10.0
Wheat germ	7.5
Water	55.9
Bulking agents	
Paper pulp (dry)	13.0
Gelgard M	1.0
Other ingredients	
Cholesterol	0.35
Choline chloride	0.07
Wesson's salt mixture	0.18
Ascorbic acid	0.8
Propionic acid	0.4
Citric acid	0.8
Total	100.0 pH 4.0 – 4.1

efforts have been concentrated on the development of a suitable mass-rearing system for the species under laboratory conditions.

The present rearing system (described in detail elsewhere: Katsoyannos and Boller, in preparation) has, at present, an output of some 20 000 eggs per day and reaches a recovery rate of about 20% (maximum 58%) from the egg to the adult stage. Because of the pupal diapause only two generations per year can be produced. This phenomenon has, on the other hand, the advantage that pupae can be produced and stockpiled throughout the year with relatively low investments with regard to rearing space and manpower.

The composition of the larval diet used at present at our laboratory is given in Table I.

2.2.   Sterilization, marking and transportation

Freshly emerged flies are transferred daily to holding containers (30-litre cylinders, polyethylene) at 24 °C, 60 - 70% RH and 15 lx on a dry diet consisting of 99.9% sugar and 0.1% Dysprosium chloride that is prepared as a thick slurry, applied to filter paper strips and dried before application. The low light intensity allows a normal feeding activity but prevents mating. The carbohydrate diet slows down the development of the ovaries and provides at the same time the marking substance for the neutron activation analysis of recaptured flies that do not show a distinct visual marker. At the age of 2 days the flies are transferred to a cold

room (0°C) and chilled, and their numbers determined by weight. Batches
of 2000 flies are processed in a simple device that distributes 5 samples of
400 flies into chilled glass bottles (20 ml volume) with a plastic snap-cover.
These bottles are transported in a cooling box to the irradiation source
(type SULZER, pool-source with eight ^{60}Co bars) and irradiated with
10 ± 0.5 krad (dose-rate 300 krad/hour). After irradiation the flies are
transferred in the cooling box to a low-temperature room (15°C), marked
with a fluorescent aerosol spray (KRYLON; Remund and Zehnder, in
preparation) and poured back into the holding containers. They are now
given the standard dry sugar-protein diet (sugar: yeast hydrolysate 4:1) and
held under low light intensity until released in the field.

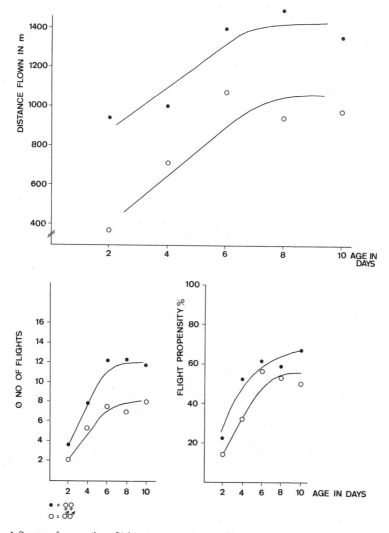

FIG. 1.   Influence of age on three flight parameters (distance flown in 24 hours, number of individual flights
and flight propensity). Averages of 50 values.

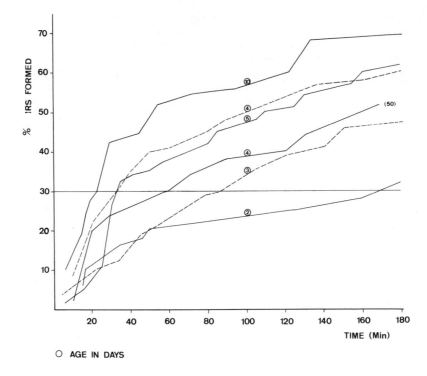

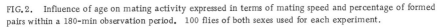

FIG. 2.   Influence of age on mating activity expressed in terms of mating speed and percentage of formed pairs within a 180-min observation period.   100 flies of both sexes used for each experiment.

Flies that have reached an age of 4 - 7 days are transported twice a week under exclusion of light in their holding containers to the release sites where the containers are opened and exposed to sunlight.   After 10 - 15 min some 90% of the flies have left their containers and flown to the cherry trees.

2.3.   Optimalizing the age of flies

Whereas a quality control programme measures the impact of rearing procedures on the individual components of the quality [7], investigations were carried out to find the optimal age of the released flies with regard to dispersal, host finding and mating activities.   Flight mill studies revealed that males and females reach their optimal flight capacity and flight propensity after 6 - 8 days (Fig. 1).   Similar observations were made with respect to mating speed as an index for the mating propensity of males and females of a given age group (Fig. 2).   These two aspects indicated that the immediate impact of the sterile flies on the wild population could be increased by holding the flies back in the laboratory for at least 4 days.   This seemed to be feasible as no extra work was involved and the holding containers required only minimal laboratory space.

3.   FIELD RELEASES

3.1.   The release sites

On the basis of experience gained in 1972 three new release sites were selected in the same general area where we had worked for several years. One orchard with 30 cherry trees was completely surrounded by dense forest, the second orchard consisted of 15 cherry trees which were separated from other orchards on two sides by dense forest and on the other sides by open fields (the distance to the closest orchard was 200 m). This latter situation was included in our experiments in order to investigate immigration and emigration of flies to and from the experimental orchard. A third site to be included in 1974 was also completely surrounded by forest and contained a mixed orchard of some 50 cherry and 50 apple and plum trees. None of the three orchards had received insecticidal treatments for several years and average infestation levels reported by the owners were medium to severe. However, due to the peculiar situation in 1972 [3] the density of the natural populations had dropped drastically because birds had removed most of the few cherries that had survived the late frost spells.

3.2.   Release schedule

On the basis of our forecasts of beginning of flight and of the peak of the cherry harvest [8] as well as on emergence data and flight curves of the target species, we released sterile flies in 3 - 4 day intervals and with increasing rates starting one week before the first flies appeared in nature and ending 10 days before harvest. During this period a total of 73 000 flies was released in both experimental orchards and monitored by seven visual traps that were checked twice a week throughout the entire flight period.

3.3.   Results

6.5% of the released flies were recaptured. More than 95% of these flies exhibited distinct fluorescent markers and the remainder had to be analysed by means of the neutron activation technique. With the exception of five doubtful cases where the autoradiographic method did not produce reliable identification of marked flies, no wild flies could be discovered. However, it was assumed that a very low wild population was present because an average number of 0.16 flies per trap was observed in one check orchard where an intensive trapping programme had been carried out (50 trees, 289 traps).

No marked flies were observed to leave the experimental orchards and to disperse over the open fields. The periodical incubation of eggs dissected from cherries showed that all eggs were fully sterile and no infestation of the crop could be detected at harvest time. In the check orchards the average infestation was 2.0% (range 0 - 8%).

4.   CONCLUSIONS

The releases carried out in the past two years have shown that the methods applied in the sterilization, marking and release of the cherry fruit fly have reached a satisfactory level and make it possible to initiate small-scale control experiments. Despite the promising results achieved in 1973 because of favourable conditions (very low population densities of the target species) it is necessary to continue the pilot studies under various ecological conditions before the feasibility of an eradication can be determined.

In addition, research has to continue in two directions: Mass-rearing (including quality control) and host-race situation of Rhagoletis cerasi. Although considerable progress has been made in the development of a suitable mass-rearing system, further improvements are necessary in order to maintain a high quality and genetic variability of the reared insects that have to replace the field-collected material used at present. Population-genetic studies in progress [9, 10] have to give us a clear answer in the near future as to whether the host-race of R. cerasi associated with honey-suckle (Lonicera spp.) should be considered a potential source for re-infestations or a distinct species that does not interbreed with the true cherry fruit flies in nature.

ACKNOWLEDGEMENTS

Many thanks are due to the staff of this laboratory — especially to V. Katsoyannos who made these studies possible. The authors also extend their thanks to Dr. K. Russ for his collaboration in the joint collection campaigns for Rhagoletis material in east Austria.

REFERENCES

[1]   BOLLER, E. et al., Economic importance of Rhagoletis cerasi L., the feasibility of genetic control and resulting research problems, Entomophaga 15 (1970) 305.

[2]   BOLLER, E.F., HAISCH, A., PROKOPY, R.J., "Sterile insect release method against Rhagoletis cerasi L.: preparatory ecological and behavioural studies", Sterility principle for Insect Control or Eradication, (Proc. Symp. Athens, 1970), IAEA, Vienna (1971) 77.

[3]   BOLLER, E.F., "Status of the sterile-insect release method against the cherry fruit fly (Rhagoletis cerasi L.) in northwest Switzerland", The Sterile-Insect Technique and its Field Applications (Proc. Panel Vienna, 1972), IAEA, Vienna (1974) 1.

[4]   PROKOPY, R.J., BOLLER, E.F., Response of European cherry fruit flies to coloured rectangles, J. Econ. Entomol. 64 (1971) 1444.

[5]   RUSS, K. et al., Development and application of visual traps for monitoring and control of populations of Rhagoletis cerasi L., Entomophaga 18 (1973) 103.

[6]   INTERNATIONAL ATOMIC ENERGY AGENCY, Radiation, Radioisotopes and Rearing Methods in the Control of Insect Pests, (Proc. Panel Tel Aviv, 1966), IAEA, Vienna (1968) 143.

[7]   BOLLER, E., Behavioral aspects of mass-rearing of insects, Entomophaga 17 (1972) 9.

[8]   REMUND, U., BOLLER, E., Neuerungen im schweizerischen Prognosewesen für die Kirschenfliege, Schweiz. Z. Obst-Weinbau 107 (1971) 183.

[9]   BOLLER, E.F., BUSH, G.L., Evidence for genetic variation in populations of the European cherry fruit fly, Rhagoletis cerasi L. (Diptera: Tephritidae) based on physiological parameters and hybridization experiments, Ent. Exp. & Appl. 17 (1974) 279.

[10] BUSH, G.L., BOLLER, E.F., Genetic distance and evolutionary relationships in the Rhagoletis cerasi group, Ent. Exp. & Appl. (submitted).

SMALL-SCALE FIELD EXPERIMENTS ON STERILE-INSECT CONTROL OF THE ONION FLY, Hylemya antiqua (Meigen)*

J. THEUNISSEN, M. LOOSJES,
J. Ph. W. NOORDINK, J. NOORLANDER,
J. TICHELER
Institute for Phytopathological Research,
Wageningen, The Netherlands

Abstract

SMALL-SCALE FIELD EXPERIMENTS ON STERILE-INSECT CONTROL OF THE ONION FLY, Hylemya antiqua (MEIGEN).

During three consecutive years small-scale field trials were carried out in order to check the feasibility of the sterile-insect technique as a possible control method for the onion fly, Hylemya antiqua (Meigen). Results, so far analysed, suggest a considerable reduction of the initial onion fly population, the possibility to attain a satisfactory level of sterile flies in the population, a normal percentage of mated sterile females, a normal susceptibility of the sterile flies to insect fungus infection and a decreasing number of infested onion plants per hectare.

INTRODUCTION

Laboratory research on the possibility to use the sterile-insect technique (SIT) to control the onion fly, Hylemya antiqua (Meigen), led to a series of four small-scale trials on onion fields of which three have been carried out. As a result of preliminary work [1, 2], research topics included: properties of the larval medium [3], pupation, production and storage of pupae, egg production and rearing efficiency [4], irradiation, competitiveness of sterilized flies and the use of radioisotopes [5], histopathological effects of irradiation [6], spermatogenesis and oogenesis [7, 8], cytogenetics [9-11], marking, dispersal, trapping methods and population dynamics [12], and computer simulation models [13-15].

Some results of the first two field trials were discussed earlier [16] and recently a general review on the genetic control of the onion fly was published [17]. As to the present situation in sterile-insect projects, Lindquist [18] published a review on the progress made during the past few years. The present paper describes some results of the first three small-scale field experiments in the Netherlands.

METHODS

During the third of the current series of small-scale field trials, fourth-generation offspring of field-collected onion fly pupae were used. The methods used have been described before [16] and only minor changes have

*This project has been supported by Euratom and the International Atomic Energy Agency (IAEA Research Contract No. 676/RB).

been made.   The pupae were irradiated with 3 kR of gamma rays from a ^{60}Co source (1.323 MeV, $T_{\frac{1}{2}}$ = 5.25 years; 120 000 Ci; dose-rate 900 R/min).

In order to mark the sterile flies in the field with a colour, the irradiated pupae were placed under a thin layer of a fluorescent powder, Radglo, in the following way.   In a tray 25 cm × 25 cm × 10 cm, in the bottom of which small holes have been drilled, a piece of filter paper is covered with a thin layer of wet sand.   About 10 000 irradiated pupae are mixed with moist vermiculite and poured on the wet sand.   The pupae are covered with a thick layer of dry sand to force the emerging pupae to use their ptilinum well, in order to pick up the stain.   The fluorescent powder is applied in a thin layer on the surface of the sand.   This vulnerable layer is protected by a thin cover of vermiculite, which is sieved on the powder, and the vermiculite is covered with a thin layer of dry sand.

Each tray was covered by an inclined plate to keep the contents dry. As a result "all weather" functioning of the tray was secured while the emerged flies could leave it freely.

Weekly, ten trays were used to release the sterilized pupae.   For six consecutive weeks a different colour was used before an already applied colour was re-used.

The release schedule was simpler than that of the two preceding field trials.   The idea of following the emergence of the wild population and covering it with a more or less constant ratio of sterile flies was abandoned in favour of the weekly release of a constant number of sterile flies.   The population development of the onion fly in the field seems to be too irregular and unpredictable to base any release system on it.   When a constant number of sterile flies is released the development of the wild population is not considered.   During the first weeks releases took place by putting two trays between platforms some 10 - 15 cm above the level of the soil of the onion field to protect the pupae from mice.   However, when weather conditions were hot, the temperature in the trays rose too much.   Moreover, this way of gradual release attracted increasing numbers of birds.   In spite of the protection measures taken, bird predation was high on some groups. Further releases took place by digging the trays in one or two groups in the soil at the margin of the onion field or putting the trays between rows of potato plants in an adjacent field.   Only the latter method stopped bird predation effectively.

In preceding years, during the first flight, the sterile flies were released on the field where the wild flies could be expected to emerge, i.e. the onion field of the year before.   During the second flight, releases were carried out on the new onion field.   In 1973, all releases took place on the present onion field to determine whether such a simplified procedure could be used or not.   Every week about 100 000 pupae, yielding an average number of 84 000 flies, were released.

Mating of the females was determined by checking the contents of the spermatheca.

In 1973, infestation of plants on both the experimental field and a check plot was checked weekly.   Certain rows were selected to cover the experimental and untreated area, representing 3.1% and 2.8% of the areas. Infested plants were marked with a label near the plant and, before the harvest, soil samples were taken where labels indicated infested plants to estimate the over-wintering population of pupae.   During the field experi-

ments the sampled areas represented the following percentages of the
experimental fields: (1970) 0.5; (1971) 0.2; (1972) 1.6; (1973) 3.1.

RESULTS

Some results of the first two field experiments have already been
discussed [16]. For comparison the years of these experiments are
included in the data presented here.

During 1970, 1971, 1972 and 1973 the output of the rearing was
increased considerably due to a steady improvement in rearing methods and
available facilities (Table I).

In 1973 the numbers of released pupae were increased significantly
compared with those of previous years. This resulted in a high level of
sterility during the experiment. The quality of the pupae and the prevailing
weather conditions permitted a high percentage of sterile flies to emerge.
As a result of the increased numbers of released sterile flies (Table II)
the percentage of sterile flies in the population was higher in 1973 than in
previous years (Fig.1). In general, a high level of sterility in the population
was obtained.

The percentage of egg sterility in 1973 was difficult to determine owing
to the low number of eggs produced. The numbers of trapped fertile females
were low and many died without ovipositing in the laboratory cages. The
limited number of egg batches used to determine the figures caused
considerable variability. Moreover, a few fertile, already mated immigrating
females could have a disproportionly negative influence on the egg sterility
(Fig.2).

During the three experiments the percentage of mated females was
scored in order to compare the mating of the sterile females with that of
the fertile ones (Fig.3). In 1973, the numbers of surviving fertile females
were often too low to draw any conclusions. As a result of fungus infestation
sometimes even sterile females did not survive in sufficient numbers.

In 1971, fungus epizootics were observed in the field (Fig.4, 1971).
The fungus was an Entomophthora sp. A marked discrepancy between the
percentages of sterile flies and sterile eggs (Figs 1 and 2) suggested that
preferential infestation of sterile flies could be a possible cause [16]. To
prevent such a presumed preferential infestation, release methods were

TABLE I. AVERAGE WEEKLY PRODUCTION OF PUPAE DURING
MASS-REARING PERIOD

1970	25 000
1971	50 000
1972	100 000
1973	200 000
1974	500 000[a]

[a] Estimated.

TABLE II.   SUMMARY OF RESULTS

	Estimated wild population of pupae in spring	Average weekly release		Mean % sterile flies in population	% infested onion plants
		Pupae	Flies[a]		
1971	20 000	27 700	25 000	56.0	5.22
1972	17 000	50 500	32 400	67.8	0.60
1973	5 000	104 800[b]	84 700	91.1	0.12
1974	400				

[a] Flies emerged in the field from the sterilized pupae.
[b] Experimental area in 1973 was 1.5 ha.

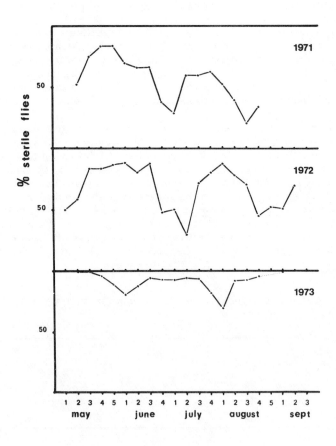

FIG. 1.   Percentages of sterile flies in population.

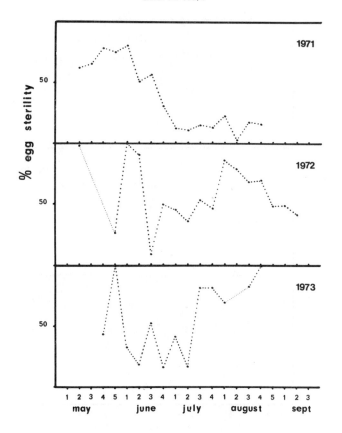

FIG. 2. Percentages of sterile eggs produced by trapped females. Figures based on 20 eggs or less have been omitted.

changed accordingly. Therefore, the rate of insect fungus infestation in the trapped, living females was recorded during the next years (Fig. 4). In 1972 and 1973, the fungus infection in trapped females did not show any preference for sterile or fertile females.

The number of infested onion plants/ha on the experimental fields decreased. The average number of onion plants per ha was about 10^6. Damage due to onion fly infestation was estimated during three years of field trials (Table II).

In 1973, a check plot was available 800 m from the experimental plot to check the effect of a population of fertile flies of the same order of magnitude without releasing sterile flies. However, previous experiments [12] have shown that the fly can cover distances up to 2000 m. Consequently, the average percentage of sterile flies in the check plot was 59%.

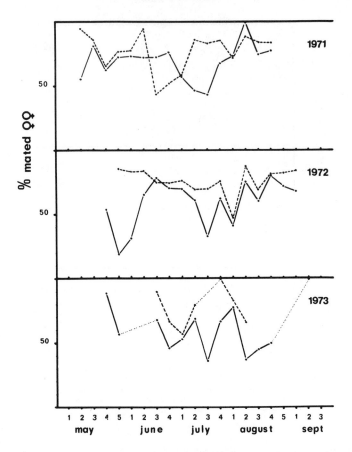

FIG. 3.   Percentages of mated females in population.   Figures based on less than 5 individuals have been omitted.

●——● sterile ♀♀   ●- - -● fertile ♀♀

DISCUSSION

As a result of the large numbers of weekly released sterile flies compared with the estimated wild population, we expected a very high sterile/fertile ratio during these experiments. In fact, this ratio proved to be lower than expected. The reasons are not clear. The flies seem to disperse outside the onion field and the attraction of an onion field seems to be limited in its ability to keep large numbers of the flies in the immediate surroundings. The apparent emigration of the sterile population is dis-proportionally disadvantageous to the effect of the sterile flies when the treated area is small. In large onion-growing areas the dispersal of sterilized flies should increase the effect of the releases.

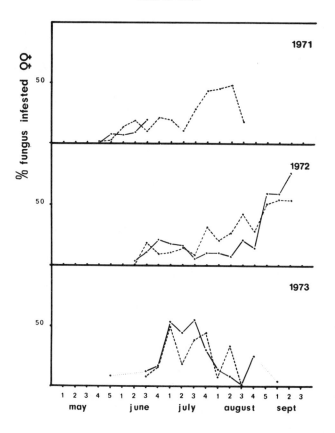

FIG. 4.   Percentages of trapped females infested with insect fungus.   Figures based on less than 5 individuals have been omitted.   Broken line in graph of 1971 represents the percentage of sterile plus fertile females.
●——● sterile ♀♀   ●--● fertile ♀♀

A possible explanation for the low egg sterility in 1973 is the immigration of already fertilized females from private gardens. When the low local population of fertile females is effectively controlled by the released steriles, a few immigrating fertilized females may have a considerable effect on the egg sterility figures. It is possible that such a low population density of wild flies was reached that a few immigrating females had a considerable influence on the total egg sterility without being important in absolute numbers. Consequently, when there is immigration an increase in the number of released sterile flies makes sense only up to a certain level when a small area is to be treated.

The percentages of fertile and sterile mated females (Fig. 3) do not seem to differ essentially. In general, the average sterile female trapped in the onion field will be younger than the average immigrating fertile female. This may be responsible for the percentage of mated sterile females being somewhat lower than that of the fertile ones.

Epizootics of insect fungus diseases doubtless play an important role in the field populations of onion flies and other Diptera. During the field trials severe outbreaks of Entomophthora epizootics were observed in the treated area. Diseased flies show a tendency to crawl towards high points in the vegetation, e.g. flowers and stalks of grass, in the last stage of pathogenesis. Infested flies may come into the traps. Before the death of the fly the fungus starts to break through the integument. At that time the insect is still able to move but its reproductive organs, intestinal tract and most of the fat and musculature have already completely disintegrated [19]. In insects, generally one of the first functions to be destroyed by an infectious disease is reproduction. So it is possible that the fly population contains many individuals that do not reproduce.

ACKNOWLEDGEMENTS

The authors would like to express their thanks to all who have contributed to our project, especially Miss E. Lemmers, Miss S.A. Voorhoeve, Mr. G. Schelling and several students of Dutch Universities.

REFERENCES

[1] FLUITER, H.J. de, NOORDINK, J.Ph.W., TICHELER, J.H.G., The onion fly problem and the sterile male technique, Meded. Reactor Centrum Nederland 20 (1967) 83.

[2] TICHELER, J., NOORDINK, J.Ph.W., "Application of the sterile-male technique on the onion fly, Hylemya antiqua (Meig.), in the Netherlands. Progress report", Radiation, Radioisotopes and Rearing Methods in the Control of Insect Pests (Proc. Panel Tel Aviv, 1966), IAEA, Vienna (1968) 111.

[3] TICHELER, J., "Rearing of the onion fly, Hylemya antiqua (Meigen)", Sterility Principle for Insect Control or Eradication (Proc. Symp. Athens, 1970), IAEA, Vienna (1971) 341.

[4] NOORLANDER, J., unpublished.

[5] NOORDINK, J.Ph.W., "Irradiation, competitiveness and the use of isotopes in sterile male studies with the onion fly, Hylemya antiqua (Meigen)", Sterility Principle for Insect Control or Eradication (Proc. Symp. Athens, 1970), IAEA, Vienna (1971) 323.

[6] THEUNISSEN, J., "Radiation pathology in Hylemya antiqua (Meigen): outlines of research", Sterility Principle for Insect Control or Eradication (Proc. Symp. Athens, 1970), IAEA, Vienna (1971) 329.

[7] THEUNISSEN, J., Egg chamber development in the onion fly, Hylemya antiqua (Meigen) (Diptera: Anthomyidae), Int. J. Insect Morphol. Embryol. 2 (1973) 87.

[8] THEUNISSEN, J., Chromatin transformations in the trophocytes of the onion fly, Hylemya antiqua (Meigen), during egg chamber development (Diptera: Anthomyidae), Int. J. Insect Morphol. Embryol. (in press).

[9] HEEMERT, C. van, Isolation of a translocation homozygote in the onion fly, Hylemya antiqua (Meigen) with a cytogenic method in combination with the determination of the percentage late embryonic lethals, Genen Phaenen 16 (1973) 17.

[10] HEEMERT, C. van, Androgenesis in the onion fly, Hylemya antiqua (Meigen), demonstrated with a chromosomal marker, Nature (London) New Biol. 246 (1973) 21.

[11] WIJNANDS-STÄB, K.J.A., HEEMERT, C. van, Radiation induced semi-sterility for genetic control purposes in the onion fly, Hylemya antiqua (Meigen). Isolation of semi-sterile stocks and their cytogenetical properties, Theor. Appl. Genet. 44 (in press).

[12] LOOSJES, M., unpublished.

[13] WIJNANDS-STÄB, K.J.A., FRISSEL, M.J., "Computer simulation for genetic control of the onion fly, Hylemya antiqua (Meigen)", Computer Models and Application of the Sterile-male Technique (Proc. Panel Vienna, 1971), IAEA, Vienna (1973) 95.

[14] FRISSEL, M.J., WIJNANDS-STÄB, K.J.A., "Computer modelling of the dynamics of insect populations",
      Computer Models and Application of the Sterile-male Technique (Proc. Panel, Vienna, 1971), IAEA,
      Vienna (1973) 23.
[15] LOOSJES, M., FRISSEL, M.J., unpublished.
[16] TICHELER, J., LOOSJES, M., NOORDINK, J.Ph.W., NOORLANDER, J., THEUNISSEN, J., "Field
      experiments with the release of sterilized onion flies Hylemya antiqua (Meig.)", The Sterile-insect
      Technique and its Field Applications (Proc. Panel Vienna, 1972), IAEA, Vienna (1974) 103.
[17] THEUNISSEN, J., LOOSJES, M., NOORDINK, J.Ph.W., NOORLANDER, J., TICHELER, J.H.G.,
      Genetic control of the onion fly, Hylemya antiqua (Meigen), in The Netherlands, Proc. 7th British
      Insecticide and Fungicide Conference, 1973 (in press).
[18] LINDQUIST, D.A., Recent advances in insect control by the sterile male technique, Mededelingen
      Fakulteit Landbouwwetenschappen, Gent 38 (1973) 627.
[19] THEUNISSEN, J., unpublished.

MASS REARING OF THE OLIVE FRUIT FLY, Dacus oleae (Gmelin), AT "DEMOCRITOS"

J. A. TSITSIPIS
Nuclear Research Center "Democritos",
Aghia Paraskevi,
Attiki, Greece

Abstract

MASS REARING OF THE OLIVE FRUIT FLY, Dacus oleae (Gmelin), AT "DEMOCRITOS".
Production of more than 4.5 million olive fruit fly pupae within a 4-month period during the summer and autumn of 1973, at an approximate cost of US $1 per 1000 pupae, was made possible by introducing certain improvements in the formerly used rearing system. Replacement of the adult liquid diet by a solid one and less frequent changing of the water supply saved labour; better timing in the egg collection improved hatchability. Incubation of the eggs in 0.3% propionic acid followed by their surface sterilization drastically cut down or eliminated previous sporadically appearing microbial contaminations. Most important, a new larval diet (T), which is much easier to prepare and handle, has doubled pupal yield. A new caging and egging system under development provides a higher egg production and requires far less labour. Preliminary promising results on new larval diets and modifications in the various steps of the rearing procedure will hopefully contribute to achieving the much lower costs needed even in a moderate-scale mass-release programme.

INTRODUCTION

Mass rearing of an insect is an unqualified term by itself. It could imply a daily production of anything between a few thousand and several million insects. Each method for rearing insects has certain inherent limitations with regard to its potential, depending mainly on available funds. Hence, reference to a method should include the capability of raising a certain number of insects under specified conditions. This paper describes an improved method of rearing up to half a million pupae per week with nine laboratory technicians and the space and equipment of the "Democritos" entomology laboratory.

The known methods for rearing the olive fruit fly have been reviewed by Orphanidis et al. (1970), who dealt solely with larval artificial diets, and by Tzanakakis (1971). Cavalloro and Girolami (1968) described a method for rearing the fly in the laboratory. The method used at "Democritos" until 1972 for mass rearing the olive fly has been described by Tzanakakis (1967), Economopoulos and Tzanakakis (1967), and Tzanakakis et al. (1968, 1970). Since then various stages of the procedure have been improved. Only these improvements are considered here. They are aimed at reducing the cost of rearing, at saving space, and at ameliorating the quality of the flies.

Also, reference is made to promising results from research underway. It is expected that the adoption of such innovations will drastically reduce the cost of rearing.

PRESENT REARING SYSTEM

The present rearing system successfully met the requirements for producing large numbers of insects for the experimental application of the sterile-insect technique in an olive grove in the Chalkidiki penninsula in northern Greece. Some 13 185 000 eggs were produced from which about 4 575 000 pupae were obtained from June 10th to October 3rd, 1973. The average egg hatchability was 66.62% (based on almost daily checks of a grand total of 10 419 eggs) and the adult emergence from pupae was above 80%. The pupal yield of the larval diet (T) was 51.42% (based on hatched eggs) or 34.25% when based on eggs laid. During the 4-month period, 6 356 trays of larval diet were prepared (approx. 1800 kg in weight).

Adult maintenance

The Hagen cage and oviposition substrates (Hagen et al., 1963) are still in use. The No. 2 liquid diet (Economopoulos and Tzanakakis, 1967), offered three times weekly to the flies, has been replaced by a solid diet developed in the laboratory (Tsitsipis, Kontos and Gordon; unpublished data) and is supplied only once at the initiation of each cage. The diet is prepared by drying the liquid diet under vacuum at 40°C. Although females feeding on liquid diet give about 20 - 30% more eggs than on solid diet (egg production expressed per living female), the higher mortality of the females on the liquid diet results in an egg production (when expressed per initial female) that is the same or less than on the solid diet. Water, supplied in a small plastic box with a wick passing through the cover, is changed every 10 - 15 days instead of 2 - 3 times weekly as was practiced before. When starting new cages, volumetrically measured pupae, rather than adults, are introduced into the cages as they are easier to handle.

To improve egg hatchability, fresher eggs are obtained by exchanging the egg-laden paraffin domes for new ones shortly before the onset of scotophase (Dr. H.T. Gordon's suggestion).

Handling of eggs and pupae

The squeeze-bottle water rinser for the collection of eggs from the domes and the slightly negative pressure applied through a Buchner funnel for filtering out the eggs have been replaced by a spray-gun rinser connected with the water mains and a funnel with a siphon and a densely woven silk fabric, respectively. The eggs are transferred with a brush onto filter paper impregnated with 0.3% propionic acid (first used by Manoukas, 1972) in Petri dishes for a 24-h incubation period at 25°C. Figure 1 indicates clearly that such a concentration is not harmful. Problems of fungal contamination of eggs that were occasionally encountered with sodium benzoate were eliminated with the use of propionic acid.

After the 24-h incubation period, the eggs are surface-sterilized with 2% Clorox ® (5 - 6% sodium hypochloride) for 10 min[1] (Fig. 2 shows the hatchability of Clorox-treated freshly laid eggs). The efficacy of sterilization was tested by aseptically seeding surface-sterilized eggs on an agar culture

---

[1] Egg hatch with Clorox treatment soon after deposition and 24 hours later was similar.

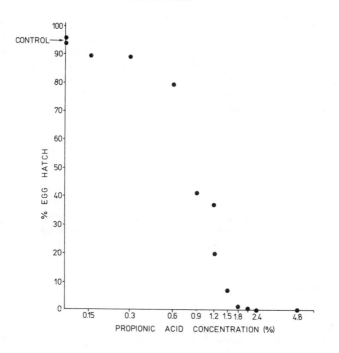

FIG. 1.   Hatchability of 0- to 6-h-old <u>Dacus oleae</u> eggs incubated in propionic acid.

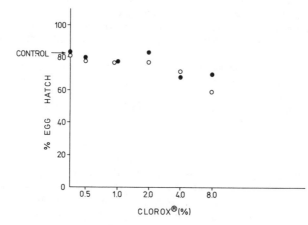

FIG. 2.   Hatchability of 0- to 6-h-old <u>Dacus oleae</u> eggs exposed on Clorox® for 5 and 10 min, rinsed with water for 10 min and incubated in 0.3% propionic acid.
● 5 min;   ○ 10 min.

medium containing enzymatic soy hydrolysate and dehydrated nutrient broth (kindly provided by Dr. Haniotakis). Although there was some larval development, there were no microbial colonies. After sterilization, the eggs are rinsed in tap water for 10 min; they are then suspended in 0.3% propionic acid and sprinkled onto the diet instead of being transferred with a brush as before. Surface sterilization has definitely decreased diet contamination with microorganisms.

The eggs are counted volumetrically on the basis of about 45 000 eggs/ml, an average figure reached after several sampling counts.

The larvae, upon completion of their development, either crawl out of the diet and pupate in the sawdust in which they drop, or pupate on or inside the diet itself (30 - 60% of them).

Upon termination of pupation, the diet is dumped into water, the remaining pupae are skimmed and then mixed with sawdust to dry. The pupae are sieved to remove the sawdust and counted volumetrically (about 900 pupae/10 ml). They are then kept in open plastic boxes and again mixed with sawdust which is removed 1 - 2 days before emergence.

Larval rearing

In an attempt to improve larval diet N (Tzanakakis et al., 1970), several modifications were made. Medium T (Table I) gave the most promising results, hence, it was used to produce the large numbers of flies required during the summer and autumn. Agar and roasted peanuts were eliminated altogether and the cellulose powder was increased by as much as 50%. The average yield of pupae per gram of diet was 2.54 (average from a total of 4 575 thousand pupae) with a range of 1 to 3.96 (averages from batches of at least 80 - 100 trays of diet), as compared with an average of 1 - 1.5 in diet N. The pupal weight was always between 5.5 - 6.5 mg, similar to that for pupae on diet N.

The preparation of diet T is much easier now that peanuts and agar are not included. Boiling and peanut grinding are no longer required and the granular consistency of the diet (similar to that used by the IAEA entomology group) makes it easier to dispense onto trays than the pasty and fast-setting diet N.

However, the granular diet requires an extra step in the just-described rearing procedure: this is the recovery of a large part (30 - 60%) of the pupae from within the diet. The quality of such pupae was found to be as good as that of the pupae outside the diet.

Cost of rearing

Tzanakakis et al. (1968) estimated a cost of US $ 13.80 per 1000 olive fruit fly pupae when the weekly fly production was 10 000 - 15 000. A current estimate at a 20-fold production rate is given in Table II. Only expendable materials and labour costs are considered. The cost estimation is based on an average daily pupal production of 64 000 pupae (90 trays × 280 g/tray × × 2.54 pupae/g), and 6.25% of the cost is allocated to replacement of terminating cages. A daily average production of 180 000 eggs obtained from 300 cages is seeded on 90 trays of diet.

TABLE I.   COMPOSITION OF SOME LARVAL DIETS FOR THE OLIVE FRUIT FLY DEVELOPED AT "DEMOCRITOS"

	N^a		T		PS^b		L^c	
Tap water	55	ml	55	ml	55	ml	100	ml
Agar (fine powder, Merck)	0.5	g	-		-		-	
Cellulose powder ($\neq$ 123, Schleicher & Schüll)	20.0	g	30.0	g	15.0 - 27.0 g^d		-	
Brewer's yeast (FIX)	7.5	g	7.5	g	7.5	g	12.0	g
Soy hydrolysate enzymatic (Nutritional Biochemicals Corp.)	3.0	g	3.0	g	3.0	g	4.0	g
Roasted peanuts	6.0	g	-		-		-	
Sucrose	2.0	g	2.0	g	-		2.75	g
Olive oil	2.0	ml	2.0	ml	2.0	ml	2.75	ml
Tween-80	0.75	ml	0.75	ml	0.7	ml	1.0	ml
Potassium sorbate (Merck)	0.05	g	0.05	g	0.05 g^e		0.05	g
Nipagin (methyl-p-hydroxybenzoate, Merck)	0.2	g	0.2	g	0.2	g^e	0.2	g
HCl, 2$\underline{N}$	3.0	ml	3.0	ml	3.0	ml	4.0	ml

[a] TZANAKAKIS, M.E., ECONOMOPOULOS, A.P., TSITSIPIS, J.A. (1970).
[b] MANOUKAS, A. (1972).
[c] MITTLER, T.E., TSITSIPIS, J.A. (1973).
[d] Solka-Floc BW-40, Brown Co., USA.
[e] Propionic acid (0.3 ml) can replace potassium sorbate and nipagin.

TABLE II.   CURRENT ESTIMATED COST OF REARING OLIVE FRUIT FLY PUPAE

	Man-hours	Cost in US $
Larval diet	-	29.26
Adult diet	-	0.08
Larval diet preparation	4.99	3.34
Adult diet	0.06	0.04
Caging, egging	24.50	16.49
Other labour	8.50	5.70
Miscellaneous (approx.)	-	5.09
Total	38.05	60.00
Cost per 1000 pupae ($ 60/60 000 pupae)		1.00

An over 90% reduction in the cost of rearing has been achieved with the introduced improvements, although it is understandable that a very strict comparison of this cost with that estimated by Tzanakakis et al. (1968) cannot be made mainly because of the different scales of operation.

TOWARD IMPROVING THE PRESENT SYSTEM

Caging and egging

As is manifest from the production costs, most of the labour (almost 65%) is devoted to maintaining the adult flies and collecting the eggs. Hence, it was deemed imperative to develop a more efficient system. Preliminary results with such a system showed that the rearing procedure could be greatly simplified. Large cages (1 m long × 0.3 m high × 0.4 m wide) are used to house the insects, and large semipermanent ceresin-coated nylon gauze cones are used as oviposition substrates. Four cones pass vertically through the cage, and the eggs that are laid on their interior walls are rinsed with water and conducted by means of a tubing system into a common receptacle. Solid diet is provided on cut-out trays on the Plexiglas ceiling of the cages. About 25 cm above the cages, 40-Watt daylight fluorescent light panels provide ample illumination for even distribution of the flies in the cages (Dr. L.F. Steiner's suggestion).

To obtain eggs with high hatchability, the cones should be changed at weekly or 10-day intervals (this is easily done), and egg collection should take place twice daily. Only 2 - 3 man-hours are required to service 12 such cages giving an approximate egg production of over 150 000. Figure 3 shows the total weekly egg production at different densities of flies per cage compared with that obtained from Hagen cages (about 1/3 as big as our new cages).

The higher egg yield and the minimal labour required suggest the adoption of this system for mass-rearing purposes, especially when anticipated further improvements, and partially automated egg collection, are introduced.

Larval diets

Table I shows the composition of two diets that appear promising for mass rearing of the olive fruit fly. Diet PS (Manoukas, 1972) is very similar to diet T. Its texture, depending on the amount of cellulose (Solka-Floc), ranges between that of diets N and T for the lower and higher amounts respectively. The pupal yield of the diet (Manoukas; unpublished results) amounted to between 2.9 and 11.0 pupae/g of diet with average pupal weights from 7.1 to 2.43 mg respectively. At 5 pupae/g, the mean pupal weight was 5.45 mg, this being the lowest weight of mass-produced pupae from diets N and T. If this diet proves successful for rearing the insects for at least 2 - 3 generations without any vigour loss, its use for mass rearing would further reduce the production cost because of the low cost of the cellulose (55 cents per kg, or about 1/5 the cost of the other cellulose used).

Diet L (Mittler and Tsitsipis, 1973) does not contain any inert material. The liquid diet is dispensed onto cotton towelling on which the eggs are placed. On a Petri-dish scale, yields of up to 20 pupae/g of diet, with a

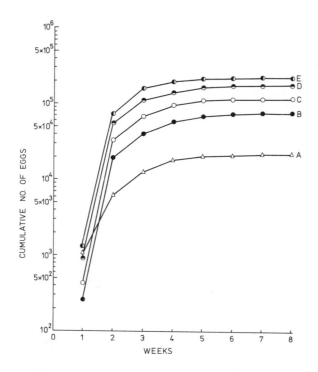

FIG. 3. Cumulative weekly egg production of <u>Dacus oleae</u> maintained at different population densities in large (B, C, D, E) and Hagen (A) cages. Each point is the mean of three replicates except for curve E where each point represents one replicate. Egg production is based on initial number of females accordingly: A = 90, B = 180, C = 370, D = 700 and E = 870.

5 mg mean pupal weight, were obtained. On a large-tray scale, microbial contamination occurred. If this problem is solved, development of a cheap larval medium is anticipated.

Two recent tests with diet T supplied in rectangular plastic trays (25 cm × 35 cm) in layers 2.2 cm thick (instead of the standard 30-cm-diameter round ones in layers 1.0 cm thick) gave pupal yields of 5 and 7.5 pupae/g of diet with a respective average pupal weight of between 5.5 and 6.5 mg. Evidently, the diets already in use are not optimally utilized, and research toward that goal is a requirement.

Having raised the efficacy of the diet manyfold, substitution of certain dietary components for cheaper ones now becomes appropriate (e.g. Schwechat brewer's yeast, kindly supplied by the IAEA entomology group at Seibersdorf, was as good as the now available FIX brand, although the former is about 8 times cheaper).

Reduction of production costs

Several steps in the rearing procedure could be simplified to lower the production costs: for example, egg incubation, egg dispensation on the diet and collection of pupae. To reduce the costs further, larger larval diet mixers and larger rectangular trays for the larval diet could be used.

Future prospects

As reported here, the production cost of 1000 pupae with the present system is about 1 US $. If the advocated improvements are incorporated in this system, we will be able to reach the 1 million pupae per week level, and the cost will hopefully drop sufficiently to compare favourably with the production costs of other insects currently mass produced, e.g. Anopheles albimanus (28 cents/1000 pupae) and screw-worm (12.5 cents/1000 pupae), although we might not reach the medfly cost of 1 cent/1000 pupae in the near future.

ACKNOWLEDGEMENTS

The author would like to express his appreciation to Drs. Economopoulos, Zervas, Haniotakis and Manoukas for their useful discussions and to Prof. Tzanakakis and Dr. Manoukas for their constructive criticism of the manuscript. Suggestions by Drs. Gordon, Steiner, Mittler and all FAO experts at "Democritos" are gratefully acknowledged.

REFERENCES

CAVALLORO, R., GIROLAMI, V. (1968), Nuove technique di allevamento in laboratorio del Dacus oleae Gmel. I., Adulti. Redia 51 p 127.

ECONOMOPOULOS, A.P., TZANAKAKIS, M.E. (1967), Egg yolk and olive juice as supplements to the yeast hydrolyzate-sucrose diet for adults of Dacus oleae, Life Sci. 6 p. 2409.

HAGEN, K.S., SANTAS, L., TSECOURAS, A. (1963), "A technique of culturing the olive fly, Dacus oleae Gmel., on synthetic media under xenic conditions", Radiation and Radioisotopes Applied to Insects of Agricultural Importance (Proc. Symp. Athens, 1963), IAEA, Vienna, p. 333.

MANOUKAS, A. (1972), Research report on nutrition and chemistry of the olive fruit fly, Dacus oleae, Nuclear Research Center "Democritos", Athens, Greece, Document No. DEMO 72/9.

MITTLER, T.E., TSITSIPIS, J.A. (1973), Economical rearing of larvae of the olive fruit fly, Dacus oleae, on a liquid diet offered on cotton towelling, Entomol. Exp. Appl. 16 p. 292.

ORPHANIDIS, P.S., PETSIKOU, N.A., PATSAKOS, P.G. (1970), Elevage du Dacus oleae (Gmel.) sur substrat artificiel, Ann. Inst. Phytop. Benaki N.S. 9 p. 147.

TZANAKAKIS, M.E. (1967), Control of the olive fruit fly, Dacus oleae (Gmelin), with radiation or chemical sterilization procedures, Nuclear Research Center "Democritos", Athens, Greece.

TZANAKAKIS, M.E. (1971), Rearing methods for the olive fruit fly Dacus oleae (Gmelin), Ann. School Agric. Forestry, Univ. Thessaloniki 14 p. 293.

TZANAKAKIS, M.E., ECONOMOPOULOS, A.P., TSITSIPIS, J.A. (1968), "Artificial rearing of the olive fly: progress report", Radiation, Radioisotopes and Rearing Methods in the Control of Insect Pests (Proc. Panel Tel Aviv, 1966), IAEA, Vienna p. 123.

TZANAKAKIS, M.E. (1970), Rearing and nutrition of the olive fruit fly. I. Improved larval diet and simple containers, J. Econ. Entomol. 63 p. 317.

COMPARATIVE BEHAVIOUR OF LAB.-CULTURED AND WILD-TYPE Dacus oleae FLIES IN THE FIELD

R.J. PROKOPY*, G.E. HANIOTAKIS,
A.P. ECONOMOPOULOS
Nuclear Research Center "Democritos",
Aghia Paraskevi,
Attiki, Greece

Abstract

COMPARATIVE BEHAVIOUR OF LAB.-CULTURED AND WILD-TYPE Dacus oleae FLIES IN THE FIELD.

Under field conditions, the authors compared the responses of lab.-type (ca. 85 generations under artificial conditions) and wild-type Dacus oleae flies to host plant colour and odour, host fruit colour and shape, small rectangles of different colours and shades, and McPhail-type traps of different colours baited with different odours. Except for the lab.-type flies being relatively more attracted toward red fruit models and small red rectangles and relatively less attracted toward yellow fruit models and small yellow rectangles than the wild type, the qualitative nature of the responses of the two fly types toward the various experimental treatments was essentially the same. Quantitatively, however, consistently smaller percentages of the released lab.-type than the released wild-type flies were recaptured, suggesting that the mobility, flight pattern, or vigour of the two types of flies may be different.

INTRODUCTION

As pointed out so well by Boller [1], a key element in the success or failure of the sterile-insect release method approach to population management is the behaviour of the released flies. As part of the current research programme at "Democritos" aimed at evaluating the potential utility of the sterile-insect release method in an integrated approach to the suppression of the olive fly, Dacus oleae (Gmelin), we initiated a series of studies comparing, under field conditions, various behavioural responses of (a) olive flies cultured under artifical conditions (yellow wax domes for oviposition, artificial media for the larvae) for ca. 85 consecutive generations and released (= lab.-released flies), (b) olive flies obtained from olives, collected in nature and released (= wild-released flies), and in most cases (c) olive flies of the natural field population (= wild population flies). None of the flies in the studies summarized here was sterilized. The particular behavioural responses we investigated included responses to host plant (= olive) colour and odour, host fruit colour and shape, small rectangles of different colours and shades, and McPhail-type traps of different colours baited with different odours. This report is a brief summary of some of our findings to date. For a detailed account see Refs [2-5].

---

* FAO-sponsored visiting scientist at "Democritos". Present address: The Farm, Rt. 1, Baileys Harbor, Wisconsin, USA.

GENERAL METHODOLOGY

Our experiments were conducted in or adjacent to unsprayed olive groves belonging to the College of Agriculture of Athens. They were located near the village of Spata, about 10 km from "Democritos".

From time of eclosion onward, the lab.- and wild-released adults were maintained under the same laboratory conditions (both sexes together in cages provided with a continuous supply of protein, sugar and water but no oviposition medium) and handled in the same manner (cooled for ca. 1 hour at 1 - 2°C for introduction into release containers). The ptilina of the flies were marked with Neon Red (lab.-released flies) or Arc Yellow (wild-released flies) Day-Glo® fluorescent powder. Marking was accomplished by placing the puparia under a thin layer of a mixture of 15% fluorescent powder and 85% talc sandwiched between two layers of sand. All captured flies were squashed between two pieces of glass and examined under u.v. light for the presence or absence of fluorescent dye. As testimony to the effectiveness of this marking method, the dye was detected in the ptilina of 99.7% of ca. 2000 lab.- and wild-released flies held in the laboratory or in a large outdoor cage for up to 60 days. Boller and Remund, in co-operation with "Democritos" (unpublished data), have evidence from flight mill studies that this particular marking technique (with 30% fluorescent powder in talc) might have detrimental effects on the flight capabilities of olive flies so marked. Whether this technique might also produce alteration of the basic nature of some of the visual or olfactory processes of the flies is unknown, but the fact that the nature of the responses of the wild-released flies conformed so closely to the nature of the responses of the wild-population flies suggests that this was not an important factor in our studies.

No actual data are presented in this paper. Instead, we have assigned the value of 100% to the experimental treatment capturing the most flies (= first-place treatment) and present the number captured in each of the other treatments as a percentage of the number captured in the first-place treatment. Where the nature of the responses of the different fly types is described as being essentially the same, the percentages given represent an average value for the combined types.

RESULTS AND DISCUSSION

Responses to colour and odour of host plant

For the sterile-insect release method to succeed, the released flies must be able to locate the natural rendezvous site for mating. Except in the case of D. cucurbitae Coquillet, the site of assembly for mating in the tephritids thus far studied is primarily, if not exclusively, confined to the host plant [6]. Although as yet we have not observed any actual matings of D. oleae in nature, we do have indications that mature flies of both sexes assemble in much larger numbers on fruiting olive trees than on non-fruiting olive trees or trees of other species. We hypothesize, therefore, that the principal mating site in D. oleae is probably fruiting olive trees. Accordingly we assessed the flies' attraction toward olive tree colour (olive tree foliage vs. fig tree foliage vs. painted surfaces vs. empty space) as well as toward olive tree odour (olive tree fruit vs. olive tree foliage and twigs vs. empty space).

For testing responses to colour, we employed two-dimensional vertically erected models of trees. To construct these models, we first built rectangular wooden frames (200 cm wide × 130 cm high, with legs 130 cm long), stretched white nylon fish net (10 mm × 10 mm mesh, each strand 0.3 mm thick) over each frame, brushed the fish net with moulten Bird Tanglefoot® to capture arriving flies, and positioned the frames in a row in an open ploughed field. We then glued fresh-picked olive or fig leaves onto 200-cm-wide × 130-cm-high rectangles of clear Plexiglas, or painted equivalent-size rectangles of plywood various colours, and placed these rectangles just behind the fish nets. We put containers of mature (14-19 days old) lab.- and wild-released flies under a dense canopy of foliated olive branches, ca. 60 cm high, 3 m from each frame. Within a few hours after each release the flies began to leave the canopy of olive foliage, those flying toward the foliage-covered or painted rectangles being intercepted and captured by the fish net. The releases were made under a canopy of host plant foliage rather than into empty space to ensure as much as possible that the flies which left the release site were not in an "escape mood". The idea of using fish net to intercept tephritid flies in flight was first proposed by Boller [7]. Indeed, to the human eye the net proved nearly invisible and captured an average of 80% of the flies which approached the rectangles.

We found that the nature of the responses of the lab.-released flies to the various experimental treatments was essentially the same as that of the wild-released flies. Both sexes of both types of flies were attracted in greatest numbers toward the bottom surface (100%) and top surface (83%) of fig leaves and the top surface of olive leaves (95%), fewer toward the bottom surface of olive leaves (60%) or a combination of the top and bottom surface of olive leaves (59%), still fewer toward any of the painted surfaces (yellow, black, red, blue, grey = 38 - 22%), and very few toward aluminium foil or empty space (6 - 7%). Additional experiments suggested that neither the odour nor pattern (two-dimensional) of fig or olive leaves nor the odour of the painted surfaces had an important influence on the responses of either type of fly. Of the 3500 flies of each type released, 8.6% of the wild-type and 7.5% of the lab.-type were recaptured on the nets.

Our conclusion is that mature flies of both the lab.- and wild-types are adept at distinguishing real foliage, apparently by its colour, from painted surfaces or empty space but that neither type is any more attracted toward olive tree foliage than toward the foliage of a non-host tree, fig.

For testing response to odour, we patterned our approach after that of Prokopy et al. [8]. Thus, we placed the odour source in wire screen containers covered with nylon tulle and hung these containers in small fruitless pear trees equipped with small sticky-coated rectangles to capture arriving flies or attached the containers to the aforementioned frames of fish net in an open field. Our release procedure was the same as that for the colour-response study except that releases were made simultaneously 3 - 12 m up-wind and down-wind from the odour source.

Fiestas Ros de Ursinos et al. [9] have reported that olive flies are positively attracted toward the odour of olives. Unfortunately, we were unable to demonstrate such attraction in our studies. Indeed, the level of response of both sexes of both the wild- and the lab.-type flies (mature, 10 days old) to washed or unwashed olive fruits (freshly picked, mixed susceptible varieties in prime stage for oviposition) was no different from that to washed or unwashed olive leaves and twigs (picked from fruiting and non-fruiting trees), a mixture of olive fruits, leaves and twigs, or the controls (= empty containers). Of the ca. 3000 flies of each type released, 4.7% of the wild-type and 3.9% of the lab.-type were recaptured. We cannot account for this lack of fly response to olive fruit odour in our studies. The weather conditions appeared ideal for this type of study and there were no competing fruiting olive trees within ca. 400 m of the pear trees or 150 m of the frames. Also, in this as well as in all our other studies, the method of release and the fact that the flies were provided with an abundant supply of food right up until release time should have ensured to a considerable degree that the flies were not in an "escape or feeding mood" when released. Perhaps the concentration of odour emanating from the ca. 15 kg of olive fruit per treatment site was too weak to elicit a positive response. Obviously, further work, possibly utilizing a different approach, is necessary before we can come to any conclusion about the comparative ability of lab.- and wild-type olive flies to detect the odour of olive fruit.

Responses to colour and shape of host fruit

In addition to being the site of oviposition, the host fruit can, in some species of tephritids, also be a specific rendezvous site for mating [10]. In an initial study we found that wild-population olive flies in olive trees flew just as frequently onto wooden models of olives as onto real olives. This finding indicated to us that once the flies have arrived on an olive tree, they locate the fruit apparently solely on the basis of its physical characteristics. We therefore assessed the responses of wild-population, wild-released, and lab.-released flies to olive-size wooden models of different shapes and colours. The models were coated with Bird Tanglefoot to capture arriving flies and were hung by wire from the branches of large olive trees (for testing responses of wild population flies) or from the branches of a small olive tree surrounded by a 2.5-m-diameter × 2.5-m-tall cylindrical saran-screen cage (for testing responses of wild-released and lab.-released flies).

We found that while the nature of the responses of all three types of flies to the various shapes of models was essentially the same, the lab.-released type responded to certain colours of models differently than the wild-types. For the colour response test, all models were olive-shaped, 31 mm tall and 20 mm in diameter at the centre. Mature females of the two wild-types were about equally attracted to yellow (102%), black (100%) and red (98%) models and less attracted to green (83%), orange (77%), blue (49%), aluminium foil (27%) and white (21%) models. However, mature females of the lab.-type were relatively more attracted to red (113%) than to black (100%) models and distinctly less attracted to yellow models (71%), with attraction to the other colours being essentially the same as that of the wild-types. Mature males of all three types of flies were most attracted to yellow, green and orange models, less attracted to black and red, and least attracted to blue, white and aluminium foil. Like the females, males of the lab.-type were relatively more attracted to red and relatively less attracted to yellow compared with wild-type males. With respect to different shapes of models (all having a surface area of 19.6 cm^2 and painted black), mature females of all three types were most attracted to olive-shaped and spherical models (100 - 99%), less attracted to cubical models (42%), and least attracted to cylindrical and rectangular models (21 - 20%). Mature males of all three types also were more attracted to olive-shaped and spherical models than

to the others, though to a lesser degree than were the females. Of the
ca. 1500 flies of each type released into the cage, 63.7% of the wild-type
and 53.1% of the lab.-type were recaptured.

Our conclusion is that females of all three types of flies are able to
detect olive fruit on the basis of its shape and colour (or contrast against
the background). The males' responses were generally similar to those of
the females, but whether they represent a response to the fruit as a rendezvous site for mating, a feeding site, or simply a response to the movements
of females captured on the models is uncertain at this time. Both sexes of
lab.-released flies were comparatively more attracted to red models and
less attracted to yellow models than the wild-types.

Responses to small rectangles of different colours and shades

Flies of a number of tephritid species are known to be more attracted
to small rectangles of yellow colour than to equivalent rectangles of any
other colour or shade [11 - 13]. In some cases (e.g. Rhagoletis cerasi L.),
small sticky-coated yellow rectangles hung in host trees have proven very
effective devices not only for monitoring fly population density and movements
but also in direct population suppression [14]. Hence, we compared the
responses of wild-population, wild-released, and lab.-released olive flies
to small (15 cm × 20 cm), Bird-Tanglefoot-coated rectangles of six different
enamel colours or shades. The rectangles were hung from the branches of
olive trees at mid-tree height, ca. 2 m apart and 50 - 100 cm from the
outermost foliage. The wild- and lab.-released flies (each type 3 - 7 days
old) were released simultaneously ca. 1 m above ground at the trunk of each
tree harbouring rectangles.

We found that the nature of the responses of the wild-released flies to
the various colours and shades tested was essentially the same as that of
the wild population flies but different from that of the lab.-released flies.
Both types of wild flies were most attracted toward the yellow rectangles
(100%), considerably less attracted toward the orange and green rectangles
(56 - 50%), and not attracted toward the red, grey and clear Plexiglas
rectangles (4 - 5%). On the other hand, the lab.-released flies were nearly
as attracted toward the orange and green rectangles (83 - 74%) as toward
the yellow (100%) and were relatively more attracted toward the red (31%),
grey (12%) and clear Plexiglas (8%) rectangles than were either of the wild
types. These colour response differences between the wild and lab.-type
flies were true for both sexes. Of the 1800 flies of each type released in
this experiment, 17.6% of the wild-type and 12.1% of the lab.-type were
recaptured. It is interesting to note that Ceratitis capitata (Wiedemann)
flies, which were also captured on the rectangles, preferred yellow (100%)
to orange and green (26 - 27%) by a considerably greater margin than any
of the three types of olive flies. Medflies were not attracted to the red, grey
or clear Plexiglas rectangles (2 - 4%).

Our conclusion is that in comparison with rectangles of green and orange
hues, yellow rectangles are more attractive to C. capitata flies (as they are
to R. pomonella and R. cerasi flies) than to olive flies. Lab.-type olive flies
proved inferior to the wild-types in hue discrimination ability. Indeed, if
grey or clear Plexiglas could be considered as neutral surfaces and used
as standards of comparison, then it is seen that the lab.-type were relatively
more attracted to red and relatively less attracted to yellow than the wild

types. Owing to the apparent lesser capability of hue discrimination in
olive flies (even wild-type) compared with some other tephritid species, we
postulate that odour-baited traps hold relatively more promise than colour-
type traps as effective devices for detecting the presence of olive flies
in an orchard.

Responses to McPhail-type traps of different colours baited with different odours

        Clear McPhail-type traps baited with different odours have been used
extensively to monitor olive fly populations in Greece. Coloured odour-
baited McPhail-type traps have also been evaluated for this purpose [15].
Hence, we compared the responses of wild-population, wild-released, and
lab.-released olive flies to McPhail-type traps of two different colours (clear
vs. traps painted daylight fluorescent yellow, the most attractive colour we
found for olive flies) baited with three different odours (water vs. aqueous
solution of 2% ammonium sulphate, the standard lure employed for detecting
olive flies in Greek olive groves vs. aqueous solution of 2% Rodia-1.5%
borax, the most attractive odour lure for olive flies found in studies at
"Democritos"), with or without a coating of Bird Tanglefoot (BT) applied to
the entire exterior of the trap. The traps were hung at mid-tree height in
the southwest parts of olive trees, one trap per tree in a ring of trees
9 - 11 m from a central olive tree (without any traps), from which the wild-
and lab.-released flies (each type mature, 8 - 15 days old) were released
simultaneously.
        We found that the nature of the responses of both sexes of all three types
of fly to the various treatments was essentially the same. All three types
were captured in greatest numbers by clear (100%) or yellow (96%) traps
baited with Rodia-borax and coated with BT. Fewer were captured by yellow,
ammonium-sulphate-baited traps with BT (62%) or clear, Rodia-borax-
baited traps without BT (55%). Still fewer were captured by yellow, water-baited
traps with BT (45%) or clear, ammonium-sulphate-baited traps with BT (32%).
The fewest were captured by clear, ammonium-sulphate-baited traps without
BT (10%) and clear, water-baited traps with BT (1%). Almost all olive flies
captured by the traps coated with BT were entangled in the BT, nearly none
being found in the interior solution, indicating that olive flies first land on
the exterior surface of a McPhail-type trap before entering the interior.
Also, the fact that clear traps with BT baited with Rodia-borax or ammonium
sulphate captured more flies than equivalent traps without BT indicates
that many flies which land on the McPhail-type traps do not enter the interior.
Of the 2100 flies of each type released in this experiment, 34.6% of the wild-
type and 20.1% of the lab.-type were recaptured.
        Our conclusion is that a clear McPhail-type trap baited with Rodia-borax
and coated with Bird Tanglefoot is an effective device for monitoring populations of all three types of olive fly studied. The addition of daylight
fluorescent yellow, the most attractive colour yet found for olive flies, does
not enhance the effectiveness of such a trap. But an odourless (i.e. water-
baited) daylight fluorescent yellow trap coated with Bird Tanglefoot is a
more effective trap than the standard clear ammonium-sulphate-baited trap,
with or without a coating of Bird Tanglefoot.

GENERAL CONCLUSIONS

For most of the behavioural reactions assayed, the qualitative nature of the responses of the lab.-cultured olive flies was essentially the same as that of the wild-types. In certain colour response assays, however, the lab.-type reacted differently from the wild-types. Specifically, the lab.-type was relatively more attracted toward red fruit models and small red rectangles and relatively less attracted toward yellow fruit models and small yellow rectangles than the wild types. We have since detected an eye colour difference between the wild- and lab.-types and have reason to suspect a deficiency or imbalance of nutrients or conditions in the artificial larval diet as being the cause of the difference in eye colour and hue discrimination ability.

On a quantitative basis, the per cent recapture of lab.-cultured released flies was in every instance smaller than the per cent recapture of wild-type released flies, supporting the earlier indication from flight mill studies (Boller and Remund in co-operation with "Democritos", unpublished data) that the mobility flight pattern, or vigour of the two types of flies may be different.

ACKNOWLEDGEMENTS

We wish to thank the College of Agriculture of Athens for making their olive groves available to us, Drs. V. Moericke, E.F. Boller and G.L. Bush for supplying us with vital equipment and supplies, and the "Democritos" staff, the FAO olive fly group at Corfu, and Drs. G. Manikas and M. E. Tzanakakis for supplying us with olive fly pupae. This study was made possible through the support of UNDP/SF. Greece 69(525): "Research on the control of olive pests and diseases in Continental Greece, Crete, and Corfu", and the Greek Atomic Energy Commission.

REFERENCES

[1] BOLLER, E.F., Behavioral aspects of mass-rearing of insects, Entomaphaga 17 (1972) 9.
[2] PROKOPY, R.J., HANIOTAKIS, G. E.,. Host detection by wild and lab-cultured olive fruit flies, Int.Symp. Insect-Host Plant Relations, Hungary (1974).
[3] PROKOPY, R.J., HANIOTAKIS, G.E., Responses of wild and lab-cultured Dacus oleae flies to host plant color, Ann. Entomol. Soc. Am. (1974).
[4] PROKOPY, R.J., ECONOMOPOULOS, A.P., McFADDEN, M.W., Attraction of wild and lab-cultured Dacus oleae flies to small rectangles of different hues, shades, and tints, Entomol. Exp. Appl. (1974).
[5] PROKOPY, R.J., ECONOMOPOULOS, A.P., Attraction of lab-cultured and wild Dacus oleae flies to sticky-coated McPhail traps of different colors and odors, Environ. Entomol. (1974).
[6] FLETCHER, B.S., ECONOMOPOULOS, A.P., Sexual behavior in fruit flies, in Report of IBP Biological Control Projects, Cambridge. Univ. Press (1974).
[7] CHAMBERS, D.L., BOLLER, E.F., FLETCHER, B.S., PROKOPY, R.J., Idea book on fruit fly movements, Document L of IBP Fruit Fly Working Group (1973) 10 p.
[8] PROKOPY, R.J., MOERICKE, V., BUSH, G.L., Attraction of apple maggot flies to odor of apples, Environ. Entomol. 2 (1973) 743.
[9] FIESTAS ROS de URSINOS, J.A., CONSTANTE, E.G., DURAN, R.M., RONCERO, A.V., Etude d'un attractif naturel pour Dacus oleae, Ann. Entomol. Soc. France 8 (1972) 179.
[10] PROKOPY, R.J., BENNETT, E.W., BUSH, G.L., Mating behavior in Rhagoletis pomonella (Diptera: Tephritidae). I. Site of assembly, Canad. Entomol. 103 (1971) 1405.

[11] PROKOPY, R.J., BOLLER, E.F., Responses of European cherry fruit flies to colored rectangles, J. Econ. Entomol. 64 (1971) 1444.

[12] PROKOPY, R.J., Response of apple maggot flies to rectangles of different colors and shades, Environ. Entomol. 1 (1972) 720.

[13] MOERICKE, V., Attraction of tephritid flies to colored rectangles, in Report of IBP Biological Control Projects, Cambridge Univ. Press. (1974).

[14] RUSS, K., BOLLER, E.F., VALLO, V., HAISCH, A., SEZER, S., Development and application of visual traps for monitoring and control of populations of Rhagoletis cerasi L., Entomophaga 18 (1973) 103.

[15] ORPHANIDIS, P.S., Contribution a l'étude du chromato-tropism du Dacus oleae (Gmel.), Experiences en laboratoire et en plein air, Z. Angew. Entomol. 74 (1973) 24.

PROGRESS REPORTS

Summaries of work done
at various institutions

SUPPRESSION OF THE MEDITERRANEAN FRUIT FLY OR MEDFLY, Ceratitis capitata (WIEDEMANN), IN A SEMI-ISOLATED AREA IN CYPRUS BY THE USE OF THE STERILE-INSECT TECHNIQUE

C. Serghiou, J. W. Balock
Agricultural Research Institute,
Ministry of Agricultural and Natural Resources,
Nicosia, Cyprus

INTRODUCTION

The Mediterranean fruit fly, Ceratitis capitata (Wiedemann), is a serious pest of citrus, peaches, apricots, figs and other hosts in Cyprus. The losses due to medfly attack and the cost of control are currently estimated at US $750 000 annually. Because of this and the geographic isolation of Cyprus, eradication of the medfly from the island is considered.

In 1972, the following preliminary operations were carried out: (a) an island-wide trapping survey, and (b) test releases of some 28 million sterile medflies from August to October. From these activities, data on wild population size and dynamics, release logistics and sterile fly dispersal were obtained. This information was used in the 1973 field experiment described here.

MATERIALS AND METHODS

Medflies were reared in the insectary at $23 \pm 2°C$ and under a 14-h photoperiod. They were kept in the usual organdy cylindrical cages loaded with 50 000 - 60 000 pupae/cage. The adult diet was a 3:1 enzymatic yeast hydrolysate : sugar mixture with water separate. Eggs laid through the organdy dropped into pans of water placed beneath the cages. The cages were kept for 15 - 17 days and produced about 6×10^6 eggs/cage.

The pupae were produced in four rooms containing each four racks of 50 larval rearing trays/rack. Each tray was loaded with 1.5 kg of medium consisting of bran, sugar, brewer's yeast, nipagin, sodium benzoate and HCl q.s. at pH4. After 9 - 10 days, each tray yielded 17 000 - 20 000 pupae.

The pupae were irradiated 24 - 48 h before emergence with 9 krad gamma rays. They were marked by mixing with a fluorescent powder (Tinopal SFG) at the rate of 1.5 g/litre pupae [1]. Before this, Day-Glo fire orange and Day-Glo rocket red, tested according to Holbrook et al. [2], were discarded because they interfered with mating.

Within the Karpass peninsula (the NE land projection of Cyprus), the villages of Rhizokarpasso and Yialoussa were selected for the experiment. Rhizokarpasso has an area of ca. 6 km² in which a variety of medfly hosts are grown mainly in home gardens; small commercial citrus groves (jaffa oranges) are also present. Yialoussa, a village of similar size and host distribution was used as control site. The two villages are separated by 18 km of forest, in which very few medfly hosts are grown.

Irradiated pupae were transported by car from Nicosia to Rhizokarpasso (85 miles; 2 h/trip) 3 - 4 times a week. To avoid overheating in transit, the pupae were packaged in 2-mil plastic bags at the rate of 2 litre of pupae/bag.

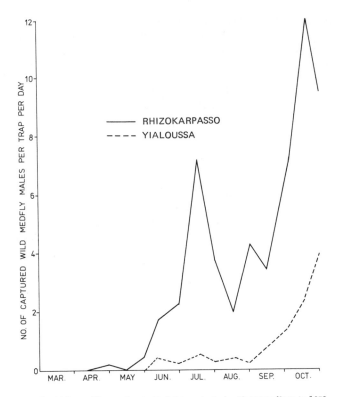

FIG.1. Male medfly catches in Nadel traps baited with Trimedlure in 1972.

The bags, closed with a rubber band, were packed in an insulated plastic picknick container together with a few freeze packs [3]. Upon arrival at Rhizokarpasso, the pupae were transferred to paper bags (50 - 300 ml/bag) with a small amount of excelsior and kept 48 h in an air-conditioned room at 23 - 27°C for emergence.

Sterile fly releases (in the late afternoon or early morning, depending on the temperature) were started in early March to coincide with low wild fly population. From early March to 10 October 1973, a total of 73 million flies (based on pupal emergence) were released. This was done by suspending on the trees the bags of emerged flies and then splitting them open. To ensure uniform distribution and minimize mating between sterile flies, the number of release points was increased from 24 to 50 and finally to a few hundred, 50 - 100 m apart.

Sixty Nadel traps baited with 2 ml Trimedlure + 3% dibrom were distributed equally between Rhizokarpasso and Yialoussa to monitor the wild fly population and the ratio of sterile to wild flies. Every week, the catches were examined under u.v. light and, in doubtful cases, by Steiner's technique [4]. Six additional traps were established concentrically some

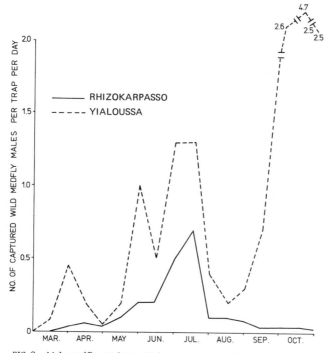

FIG.2. Male medfly catches in Nadel traps baited with Trimedlure in 1973.

2 km beyond the periphery of Rhizokarpasso. Medfly survival at Rhizokarpasso and at Deftera, a village near Nicosia, was recorded during the
release programme. Fruit collected in Rhizokarpasso and Yialoussa as it
became available was counted, weighed and held in cabinets to determine
the number of larvae developing.

RESULTS AND CONCLUSIONS

Figures 1 and 2 showing the wild fly catches in Rhizokarpasso and
Yialoussa for 1972 and 1973, respectively, illustrate medfly suppression
in Rhizokarpasso as a result of the 1973 releases. The Rhizokarpasso
peripheral traps captured 45 marked and 17 wild flies during 1973.

At Deftera, maximum survival was 57 days; however, as evidenced
by trap catches, the half-life of the released flies was no more than 1 week.
Similar data were obtained at Rhizokarpasso. In 1972 and 1973, 1 week after
suspension of releases, trap catches dropped to about one half and 2 weeks
later to about one quarter. From the third week onwards, the drop was
much more pronounced. These data suggest that the absence of a susceptible
host for 2 - 3 months would be sufficient to break the medfly cycle.

Data on larval infestation in sour and jaffa oranges, apricots and figs
are presented in Table I. The most dramatic differences in infestations

TABLE I. COMPARATIVE DATA FOR INFESTATION OF SOUR ORANGES, JAFFA ORANGES, APRICOTS AND FIGS AT RHIZOKARPASSO AND YIALOUSSA, 1973

| | Rhizokarpasso | | | | | Yialoussa | | | | |
| | | | | Infestation | | | | | Infestation | |
Date	No. of fruit	Wt (kg)	No. of larvae	per fruit	per kg	No. of fruit	Wt (kg)	No. of larvae	per fruit	per kg
Sour oranges										
2 Mar.	-	-	-	-	-	63	4.670	290	4.60	62.10
8	110	17.800	24	0.22	1.35	155	17.610	152	1.00	8.63
31	100	7.800	1	0.01	0.13	100	7.340	97	0.97	13.22
7 Apr.	100	10.100	10	0.1	1.0	100	10.500	528	5.28	50.29
21	100	10.200	53	0.5	5.2	100	11.200	1224	12.24	109.29
2 May	200	18.600	6	0.03	0.3	100	9.400	1021	10.21	108.61
11	200	22.500	3	0.02	0.1	66	5.800	157	2.38	27.07
19	53	9.200	31	0.6	3.4	38	0.900	116	3.05	128.89
26	82	11.200	26	0.3	2.3	68	7.400	131	1.93	17.70
2 Jun.	84	11.740	57	0.7	4.9	43	4.200	106	2.47	25.24
16	27	3.200	1	0.04	0.3	-	-	-	-	-
14 Jul.	51	8.600	32	0.6	3.7	-	-	-	-	-
20	14	1.305	18	1.3	13.8	-	-	-	-	-
	1121	132.245	262	0.23	1.98	833	79.020	3822	4.59	48.38
Jaffa oranges										
2 Mar.	-	-	-	-	-	4	0.445	1	0.25	2.25
8	72	9.475	0	0	0	99	13.630	1	0.01	0.07
31	100	11.785	0	0	0	100	12.500	0	0	0
7 Apr.	100	13.000	0	0	0	100	11.050	35	0.35	3.17
21	100	13.000	0	0	0	100	12.000	74	0.74	6.17
	372	47.260	0	0	0	403	49.625	111	0.28	2.24
Apricots										
2 Jun.	306	3.720	4	0.01	1.07	-	-	-	-	-
8	616	6.860	85	0.14	12.39	403	7.972	501	1.24	62.8
13	348	4.200	39	0.11	9.29	392	5.500	560	1.43	101.8
18	1190	12.900	42	0.04	3.26	-	-	-	-	-
26	139	1.053	6	0.04	5.70	-	-	-	-	-
14 Jul.	210	2.750	23	0.11	8.36	-	-	-	-	-
	2809	31.483	199	0.07	6.32	795	13.472	1061	1.33	78.8

TABLE I.   (Cont.)

Date	Rhizokarpasso					Yialoussa				
	No. of fruit	Wt (kg)	No. of larvae	Infestation per fruit	per kg	No. of fruit	Wt (kg)	No. of larvae	Infestation per fruit	per kg
Figs										
18 Jul.	100	3.200	15	0.15	4.69	100	2.150	190	1.90	88.4
4 Aug.	200	6.900	4	0.02	0.58	200	6.000	482	2.41	80.3
11	340	12.350	40	0.12	3.24	296	7.650	122	0.41	15.9
18	324	9.500	2	0.01	0.21	315	7.150	20	0.06	2.8
25	307	9.910	8	0.03	0.81	302	5.520	0	0	0
1 Sep.	300	5.905	2	0.01	0.34	280	6.430	24	0.09	3.7
8	291	5.610	4	0.01	0.71	300	5.575	32	0.11	5.7
15	300	5.020	7	0.02	1.40	300	6.180	308	1.03	49.8
22	300	4.310	7	0.02	1.62	300	6.105	1001	3.33	164
29	314	4.800	0	0.0	0.0	313	5.570	1010	3.23	181
6 Oct.	300	4.625	12	0.04	2.59	300	4.920	1619	5.40	329
13	342	5.660	8	0.02	1.41	300	6.285	854	2.85	136
20	300	6.100	3	0.01	0.49	300	6.250	1287	4.29	206
	3718	83.890	112	0.03	1.33	3606	75.785	6949	1.93	91.7

between the two villages were obtained in figs in the fall and matched the differences obtained in comparative data for wild fly traps during the same period.

During the releases, infertile or empty stings by irradiated females were observed in stone fruits during summer and in jaffa oranges in early October, when these fruits begin to change colour. These stings should be taken into account but should not deter from implementing programmes of sterile medfly releases because of the large immediate and long-range benefits of such programmes.

REFERENCES

[1]   SCHROEDER, W.J., CUNNINGHAM, R.T., MIYABARA, R.Y., FARIAS, G.J., A fluorescent compound for marking tephritidae, J. Econ. Entomol. 65 (1972) 1217.
[2]   HOLBROOK, F.R., STEINER, L.F., FUJIMOTO, M.S., Mating competitiveness of Mediterranean fruit flies marked with fluorescent powders, J. Econ. Entomol. 63 (1970) 454.
[3]   TANAKA, N., Personal communication.
[4]   STEINER, L.F., A rapid method for identifying dye-marked fruit flies, J. Econ. Entomol. 58 (1965) 374.

STATUS AND PROSPECTS OF MEDITERRANEAN FRUIT FLY CONTROL BY THE STERILE-INSECT TECHNIQUE IN PERU

J. E. Simón F
Dirección General de Investigaciones Agropecuarias,
Ministerio de Agricultura,
Lima, Peru

In Peru, mass reared and sterilized medflies were released during 1972 and 1973 as part of an integrated control programme. The release site was the Moquegua valley which contains various fruit-bearing hosts. It starts at 2 800 m in the Andes and extends to the west down to the sea for about 80 km. The valley is isolated north and south by barren hills and a 40 - 50 km desert. From sea level to the east, the experimental area was divided according to treatments into three successive zones: (1) a 40-km-long control zone where trapping only was carried out, (2) a 20- to 25-km-long release zone with some insecticide spraying and (3) the insecticide zone.

Flies were produced and sterilized with 9 krad gamma rays at La Molina, 1 200 km north of Moquegua, in provisional facilities, allowing for a maximum production of 40 million flies/week. However, a unique "medfly factory" is being completed. Its facilities provide: (1) a 15 hp steam boiler for disinfecting equipment and supplies, (2) a laundry, (3) a monorail system for transporting cages etc., and (4) a close-circuit television system to enable the superintendent to control 80 workers. Warehouse space in the basement will stock 6 months' supplies. The factory includes a 1 600 m^3 air-conditioned adult room; a 1 300 m^3 larval jumping room; a 300 m^3 pupation room; two pupae holding rooms of 200 m^3 each; space for quality control, offices and a cafeteria. The production will reach 250 million flies/week.

Within the three zones, 275 Steiner traps baited with Trimedlure were set up. Between October 1972 and September 1973, the wild population reached its peak in the control zone in December, whereas in the other two zones it went through a maximum in May. In addition to trapping, infestation was followed by fruit sampling in the three zones.

Until July, the cumulative number of chemical treatments per tree was greater in the release zone (6.45) than in the insecticide zone (6.30); however, by September, the total for the release zone (6.74) was less than for the insecticide zone (7.46). In the release zone, some 530.5 million sterile flies were released with a monthly average of 44.2 (range: 10.0 - 69.6).

The results were not clear-cut, e.g. infestation of guavas (a favourite host of the medfly) harvested in April to September was 18% in the control zone and 19 - 74% in the release zone, but mandarines harvested from May to August were 20% infested in the control zone and infestation-free in the release zone where the population peak is in May. The period of this report was fraught with technical difficulties. These are being solved and the trials will be repeated with the larger medfly facility in operation.

STATUS AND PROSPECTS OF MEDITERRANEAN
FRUIT FLY CONTROL INVESTIGATIONS BY THE
STERILE-INSECT TECHNIQUE IN ARGENTINIA

A. Turica
Instituto Nacional de Tecnología Agropecuaria,
C I C A,
Castelar, Buenos Aires,
Argentina

The annual fruit production of Argentina is some 4.2 million tons of
which 2.5 are grapes, 0.2 stone fruits and 1.5 citrus fruits. Of this total
annual production, 10 - 20% are damaged by the medfly and only 2% by the
South American fruit fly (Anastrepha fraterculus). An ecological study carried
out in northern Argentina showed that A. fraterculus is active during the
rainy season (November-March) whereas the medfly is active during 9 months
(October-June). Hence, the much greater importance of the medfly as a
citrus pest.

Between 1966 and 1973, three field experiments were carried out to test
the feasibility of the sterile-insect technique for medfly control.

The first two experiments took place in the Concordia province (just
north of Buenos Aires). In 1967, a 30-ha citrus area with a history of 40%
infestation in grapefruits was the release site. A screen of Trimedlure
traps at a density of 7/ha was installed to decrease the initial wild male
population. Subsequently, 1.275 million medflies sterilized with 10 krad
were released during February to May. From March to July, the trapped
population dwindled to zero. During the following 3 years no further medflies
were trapped in the area, nor was there any fruit infestation.

The second experiment took place between 1968 and 1969 in a 2 700-ha
semi-isolated commercial citrus area also containing some peaches and
figs. Bait sprays preceded the releases of sterile medflies during the
summer of 1968. A total of 5 million such flies were released on a weekly
schedule. The number of trapped medflies decreased and by April 1969
only 25 were caught in the release area. As opposed to this, in the chemically treated control area, the catch was then 110 - 116 flies. Grapefruit
infestation in the release area was 0.4% compared with 15% in the control area.

The third experiment was initiated during 1973 in a 60-ha isolated
citrus area in the north-western part of the country (between Chile and
Bolivia). The effect of trap density on male catch was studied: in 3 months,
2.0 Trimedlure traps/ha caught 202 medflies whereas with 0.15 traps/ha
the catch for the same period was nil. The medfly population was thus first
reduced by bait sprays and Trimedlure traps at a density of 2.0/ha. The
results of subsequent releases of sterile medflies will be evaluated by
fruit sampling, trap catches of live females and tests of their fertility in
the laboratory.

In addition, two small field tests were made in the Buenos Aires
province in a 2-ha orchard of mixed species. The flies released were
irradiated in the late pupal stage with, respectively, 7 krad and 9 krad
gamma rays. They were given a protein hydrolysate diet to increase their

survival. On an average, 3.5 million irradiated flies were released during
4 months and the area received no other treatment. At harvest, the mean
infestation of citrus fruit in the release area was 1% compared with 18%
in the untreated control area. There was some sting damage to grapefruit
in the release area.

STUDIES ON THE CONTROL OF THE MEDITERRANEAN FRUIT FLY,
Ceratitis capitata (WIEDEMANN), USING GAMMA RADIATION

A. M. Wakid
Atomic Energy Establishment,
Cairo

A. Shoukry
National Research Centre,
Dokki, Cairo, Egypt

The research described had two objectives: (a) to improve the larval diet
used hitherto in Egypt for rearing the medfly and (b) to test the competitiveness of irradiated males released in field cages together with similarly
treated females and normal individuals of both sexes.
    In improving the larval rearing medium, the aim was to decrease the
cost of laboratory-scale production. This was done by replacing dehydrated
carrot powder by wheat bran. The composition of the larval diet used
was (in g or ml, respectively): wheat bran 1 000; brewer's yeast 250;
sucrose 250; HCl (2N) 15; sodium benzoate 15; water 2 000. The cost of
1 000 pupae produced on this medium is 0.02 E£ compared with 0.04 E£ with
the carrot medium; also, the mean pupal weight increased from 8.0 mg to
9.8 mg. Adults produced were kept in the usual cylindrical hanging cages
made of synthetic cloth with plastic dishes at either end. Flies were fed a
mixture of yeast hydrolysate: sugar (1:3 parts) put in a Petri dish on the
bottom of the cage. Drinking water was offered in a glass jar hanging from
the roof of the cage. Eggs laid through the cloth fell into water within a
tray of wider diameter than the cage and put below it. Six field cages were
used in the competitiveness tests. They comprised two groups of three
adjoining cages separated by a service corridor; the outward facing aspects
were of wire screen, the other partitions glass. Each cage measured
3 m × 3 m × 2.5 m. A Casuarina equisetifolia tree within the cages was used
for hanging three oviposition devices. These were plastic mandarines half
filled with water with the upper half finely perforated. Food and water for
the adults were provided.
    The test flies had all been reared on the medium described. Pupae
and adults were kept at 25 ± 2°C; the pupae were irradiated 1 day before
emergence in a gammacell delivering 45 rad/s. In the tests, known volumes
of pupae were used from which samples were taken to determine %
emergence. In this manner, the number of flies released and the ratios of
irradiated-to-normal flies could be estimated. A control cage of 30 ml

TABLE I.   EGG HATCH OBTAINED WHEN MALE AND FEMALE
Ceratitis capitata IRRADIATED (I) ONE DAY BEFORE EMERGENCE
WERE COMBINED WITH UNTREATED (N) MALES AND FEMALES IN
VARIOUS RATIOS, TOGETHER WITH THE EXPECTED EGG HATCH AND
MALE COMPETITIVENESS VALUES

Dose (krad)	Test ratio of: $I\male : I\female : N\male : N\female$	Avg. % observed egg hatch	Avg. % expected egg hatch	Avg. competitiveness value	Avg. weekly competitiveness value		
					1st week	2nd week	3rd week
5	0 : 1 : 1 : 1	94.3	-	-	-	-	-
	1 : 0 : 0 : 1	5.4	-	-	-	-	-
	2.1 : 2.1 : 1 : 1	80.2	34.08	0.089	0.20	0.07	0
	12.9 : 12.9 : 1 : 1	50.3	11.79	0.076	0.15	0.07	0
7	0 : 0 : 1 : 1	92.9	-	-	-	-	-
	1 : 0 : 0 : 1	2.3	-	-	-	-	-
	1.8 : 1.8 : 1 : 1	82.1	34.66	0.075	0.18	0.04	0
	13.5 : 13.5 : 1 : 1	50.1	8.55	0.067	0.12	0.08	0
9	0 : 0 : 1 : 1	90.2	-	-	-	-	-
	1 : 0 : 0 : 1	1.1	-	-	-	-	-
	1.9 : 1.9 : 1 : 1	80.6	31.82	0.063	0.15	0.05	0
	12.6 : 12.6 : 1 : 1	51.2	7.65	0.062	0.14	0.07	0

normal pupae served for testing the fertility of untreated males.   The
sterility of the treated males was assessed in the laboratory by confining
25 irradiated males with 25 normal females and testing egg hatch during
3 weeks.

About 1 500 adults/cage were used and egg samples were taken for
incubation 5 times/week until the females died or stopped ovipositing.
Competitiveness of the males released was assessed by using Fried's
parameter[1]. Results are shown in Table I.

Doses of 5 - 9 krad led to almost similar reduction in egg hatch but
competitiveness decreased with increasing dose. At all doses tested, the
competitiveness of the males was less at 13 : 13 : 1 : 1 ratio then at 2 : 2 : 1 : 1
ratio. However, there was no clear indication of whether male competitiveness at a particular dose was affected by the ratio of irradiated males to
untreated males and females.

At both ratios, 5-krad-treated males gave the best competitiveness
values. The values were higher than those obtained with males treated with
7 or 9 krad. The 9-krad-treated males gave the poorest control. This may
be attributed to the adverse effect of that dose on male competitiveness.

---

[1] FRIED, M., Determination of sterile insect competitiveness, J. Econ. Entomol., 64 (1971) 869.

Generally, the competitiveness of males decreased in successive weeks, but the rate of decline was greater at the 2 : 2 : 1 : 1 ratio than at the 13 : 13 : 1 : 1 ratio. In the third week, egg hatch was as high as the controls (normal males with normal females). This may be attributed to the short life of the treated males.

RAISING AND MULTIPLYING OF THE EUROPEAN CHERRY FRUIT FLY, Rhagoletis cerasi L.

A. Haisch
Bavarian State Institute for Soil and Plant Cultivation,
Munich, Federal Republic of Germany

Availability of large numbers of cherry fruit flies is a prerequisite for controlling this pest by the sterile-insect technique. Therefore, considerable efforts were spent in developing and improving an artificial larval diet giving high yields of pupae in the shortest possible time. Another problem was to determine the conditions for shortening the post-diapause development of the pupae. The following is a summary of the work carried out to fulfill these objectives.

The artificial larval diet developed contained (in g or ml, respectively): Torula yeast 5. 6; wheat germ 4. 0; sucrose 4. 0; Vanderzant's vitamin mixture 1. 0; agar-agar 4. 2; propionic acid 0. 4; HCl (4$\underline{N}$) 0. 4; water 80. 4. The final pH of this medium was 4. 0. This larval diet, dispensed in Petri dishes, was used as a standard for the nutrition experiments. First-instar larvae, gathered 3 times a day from eggs collected daily and incubated, were seeded individually on this diet. Under these conditions, pupation occurred within 9 - 14 days, pupal yield was 46% with a mean pupal weight of 3. 9 mg and the medium remained free from overt microbial contamination. However, when eggs were collected at longer intervals and first instars transferred in groups of 20 - 50, larval development took 3 - 4 weeks, pupal yield dropped to 10%, pupal weight to less than 3 mg and 50% of the pupae died after 3 - 5 weeks. These effects are ascribed to bacterial contamination favoured by less frequent collections of eggs and neonates.

Torula- and brewer's yeast were compared as larval nutrients. There was a highly significant difference in pupal mortality: 50% with torula yeast and 15 - 34% with brewer's yeast. Torula imparted a "greasy" consistency to the medium.

When the amount of wheat germ in the diet was varied (0 - 3. 6% by weight) the optimum concentration for pupal weight and yield was 3%. To determine the nutritive role of wheat germ, this ingredient was successively extracted with six solvents of increasing polarity. The successive extracts, the residues and their corresponding combinations were added to the standard diet in replacement of 4% (by weight) wheat germ. No pupation was obtained with any of the extracts; however, all the residues and successive combinations of residue and extract resulted in pupal production. Only with the last two solvents in the series (methanol and methanol and water), which extract protein, was there a significant drop in percentage pupal yield. The evidence suggests that wheat germ has more of a textural then a nutritive function and thus stresses the importance of the physical condition of the diet.

The effects of storing pupae for different periods of time and at different temperatures on the duration of pupal diapause were studied. Batches of pupae from the Federal Republic of Germany and Hungary were kept each at 0°, 2°, 4° and 6°C for 120, 140 and 180 days. The parameters measured were: (a) the % emergence, (b) the emergence period, and (c) the mean post-diapause development period in days. (PDD, i.e. time elapsed between end of storage and mean period of pupal emergence.)

The allopatric pupal populations differed most in their PDD: low temperatures or a short storage period prolonged the development of Hungarian more than German pupae.

Between 4° and 6°C, diapause metabolism was observed to proceed fastest. Therefore, in practice, pupae should be stored at these temperatures for 120 - 140 days to avoid a drop in % emergence or a prolongation of the PDD.

PERFORMANCE OF NORMAL AND GAMMA-RAY STERILIZED
LABORATORY-REARED Dacus oleae FLIES IN THE FIELD.
AN ATTEMPT TO SUPPRESS THE NATIVE POPULATION IN A
SEMI-ISOLATED AREA BY THE STERILE-INSECT TECHNIQUE

A.P. Economopoulos, G. Haniotakis, N. Avtzis, J. Tsitsipis, G. Zervas,
A. Manoukas
Nuclear Research Center "Democritos",
Aghia Paraskevi, Attiki, Greece

In Greece, the olive fly Dacus oleae is the most serious pest of the olive fruit. Research is being conducted at the Democritos Nuclear Research Center with the objective of developing control of the olive fly by the sterile-insect technique as one of several integrated strategies. From June 1971 to November 1973, a series of field tests were carried out which laid the basis for and culminated into a trial of olive fly suppression by the sterile-insect technique. Both these tests and the control experiment took place in a semi-isolated olive grove in northern Greece (Kassandra, Halkidiki Peninsula); they are summarized below.

The first series of release/recapture tests (June 1971 - July 1972) dealt with point releases of small numbers (maximum 3 500) of olive flies artificially reared and sterilized in the pupal stage. The releases took place at the end of the spring, beginning and end of summer, early and late fall and mid-winter. Trapping results revealed a tendency of the wild flies to concentrate at the released site. By the beginning of June very few wild females had mated or were gravid whereas at the beginning of the fall, the majority contained spermatozoa and mature eggs. By the end of the fall, insemination and gravidity declined and reached a minimum in winter. Winter captures took place on warm sunny days which indicates that, in northern Greece, the olive fly can overwinter as an adult. In general, released flies were recovered mainly during the following 2 weeks.

In the second release/recapture test series (March 1972 - April 1973), large numbers of adults (maximum 41 092) were used. Six single-point releases were made with both sterilized and normal flies (five in spring, summer, fall, one in winter) and one 4-point release with sterilized flies only (winter). All flies were reared on an artificial diet except for the normal flies used in the winter single-point release: these were reared in olives.

All sterile flies were irradiated in the pupal stage, 1 - 2 days before emergence. The pupae were transported by car on the day of irradiation and kept below 28 - 30°C. They were put in trays and covered by successive layers of (a) sand (1 mm thick); (b) 30% fluorescent dye in talc (1 - 2 mm) and (c) sand (8 - 10 mm). These trays were placed in emergence cages containing adult food. Sterilized and normal flies were released from the cages simultaneously and the food was then removed. The trapping screen was not baited during 1 - 5 days following release to assure free dispersal. Thereafter the traps were baited with 3% aq. "Rhodia" protein (preservation later improved by adding 1.5% borax). Again wild flies were strongly attracted to the release site possibly because of a high concentration of pheromone(s). In October − November, the decline in insemination and gravidity among wild females was again observed. Most of the released flies were found less than 60 m from the release points and, therefore, in a control operation these should not be more than 100 - 120 m apart. Likewise, since the ratio of released/wild flies within 60 m of the release point remained high only during 8 - 10 days, weekly releases would be required.

In the winter single-point releases of normal flies reared on olives and sterilized flies reared artificially, the saliant result was the recovery of a normal fly 4 months after release, again confirming the possibility of adults overwintering in northern Greece.

As a last preliminary to the control experiment, artificially reared flies were irradiated as adults and not pupae. Before their release, they were kept in bags for 24 h either active or immobilized by cooling. Sucrose was provided. The main advantages of this procedure were found to be very low mortality, good dispersal and up to 40% recovery.

The preceding information was used in the control experiment carried out during May − November 1973. Starting during the latter part of May, 800 trees were sprayed with Lebaycid, followed in mid-June by Dimecron with "Ceratene" protein. Two weeks after the second treatment, weekly releases were started and continued until November 1973. Each of these consisted of 100 000 - 150 000 adults of both sexes treated with 9.5 krad gamma rays when 1 - 4 days old. The adults were handled as in the last test and transported by night. They were released in the morning from points less than 60 m apart. Each week, olives were examined for infestation and live trapped flies for sterility. Trapping and fruit infestation records from two untreated groves, respectively 2.7 km and 3.3 km distant from the release grove, provided comparative data. By October − November, the overall infestation in the release grove was 3 times lower than in the control groves. However, in the release area fungus infestation of the fruit was 3 times higher than in the control groves; this is probably due to stinging by released females.

RELEASE OF STERILE AND MARKED OLIVE FLIES
ON THE ISLET OF SIT (KORNAT ARCHIPELAGO)

D. Brnetic

Institute for Adriatic Agriculture and Karst Reclamation,

Split, Yugoslavia

In Yugoslavia, some 5 million olive trees are distributed almost along the entire coastal territory including the islands. During the past 10 years, interest in this crop has revived because of the canning industry. The olive fly, Dacus oleae (Gmelin), destroys about 35% of the olives annually. Control of this pest is carried out with organic phosphorus insecticides which cause residue problems and, more importantly, a calamitous upsurge of the scale Saissetia oleae. Therefore, an effort is at present being made to develop biological control of the olive fly, including the sterile-insect technique. In the field, this effort centres on an ecological study of the olive fly carried out on some islets within the Kornat Archipelago located off Split along a SE-NW direction. These rocky islands are uninhabited; on them, some pockets of arable soil with 50 to a few hundred olive trees form oases cultivated by owners living on the continent. The climate is mild with mean temperatures increasing from 7.1°C in January to 24.1°C in August and down again to 8.6°C in December. The average RH is 65% and snow is very rare.

McPhail traps baited with 2% aq. ammonium bicarbonate were established on the Kornat islands Sit and Sćitna, screened from the mainland by Pasman island on which traps were also located. In addition, traps were put on the island of Gangaro windward and off the southern tip of Pasman.

The traps were serviced in conjunction with the releases of colour-marked and sterilized flies received from Seibersdorf. Wild females caught were dissected to assess gravidity. Samples of olives were examined periodically to follow natural infestation.

During 1972 and 1973, wild flies were trapped in winter as well as in the warm season, the population reaching a maximum in September. Very few wild females trapped in June contained mature eggs. However, gravidity rose steadily during July and the beginning of August, subsided during the latter part of August to the beginning of September, reached the season's peak in September and decreased thereafter. Fruit infestation reflected this trend: in September to October infestation reached and exceeded 100%, i.e. there was more than one individual in a single fruit. The same trend was observed in 1973 when infestation again began to rise at the beginning of autumn.

During May — October 1973, an average of 3 343 flies were released on a bi-weekly schedule. This was done from a single point on Sit where marked flies were recaptured by the furthest trap, i.e. 1 km away from the release site. However, marked flies were also recovered as far as Pasman, i.e. 4.5 - 5.0 km downwind. There is little doubt that this dispersal was due to the wind because, apart from the traps on Sit and Pasman, none of those on Sćitna and Gangaro (respectively 3.0 and 12.0 km from the release point) caught any marked flies.

This work will be repeated with larger releases.

CONTROL OF FRUIT FLIES BY THE STERILE-INSECT TECHNIQUE.
STUDIES ON STERILIZATION, POPULATION DYNAMICS, FRUIT
INFESTATION SEQUENCE AND DISPERSAL OF Dacus dorsalis HENDEL

R. S. Rejesus
University of the Philippines at Los Baños,
Laguna, Philippines

In the Philippines, exportation of mango, water-melon, cantaloupe, etc.,
is limited by the oriental fruit fly, Dacus dorsalis Hendel. Control of this
pest is envisaged as a combination of the sterile-insect technique, male
annihilation, bait spraying and post-harvest commodity treatments. The
various studies required for the implementation of this programme are
divided among three national institutions: the Atomic Energy Commission,
the Bureau of Plant Industry and the University of the Philippines
at Los Baños.

The biology of the oriental fruit fly was studied on banana, mango,
papaya and guava: there was no significant difference in the duration of the
developmental stages. At 80 to 85°F the total life cycle took 19.28 days,
pre-oviposition lasted an average of 8 days and adult longevity extended from
20 to 25 days.

The fly prefers fully ripe fruits for oviposition but can also use less
mature hosts. All 54 host species tested supported development up to the pupal
stage, but there were differences in larval development, size and mortality.

For mass rearing, a larval medium based on locally available nutrients
was developed containing in grams: cooked yellow sweet potato 50; rice
bran 100; a local brand of brewer's yeast (SMC) 33; sugar 10; Na benzoate
1.5; water 250; HCl q.s. at pH 4.5. The incorporation of yeast shortened
larval development to 6.5 - 7.0 days and pre-oviposition to 6.2 - 7.7 days.
With this diet, a million pupae cost US$ 25 - 30. The medium was dispensed
in trays within muslin cloth cabinets. The eggs were seeded onto strips of paper
towelling laid on the surface of the medium and covered with newspaper. Larvae
8 - 9 days old were washed out of the medium and put in moist sawdust for pupation.

Stock cages contained cotton bolls saturated with a mixture of sugar,
yeast hydrolysate and water. Plastic cups perforated with a needle served
as oviposition substrates from which the eggs were brushed or rinsed out.

Population dynamics and dispersal were studied by trapping in a semi-
isolated 150-ha mango plantation. Before deciding on the trap design to be
used, the Steiner, Nadel, Sticky Wing and a locally designed trap were
compared in a 20-ha mango orchard. The local trap is a cylinder (diameter
3.5 in; length 6 in) of screen wire (0.5 in) with both ends fitted with semi-
circles of the same wire: suspended in its middle is a dental roll baited
with methyl eugenol/insecticide. The cylinder is placed inside a perforated
polyethylene bag which is then suspended horizontally. The insects trapped
can be collected by simply replacing the bag. This trap, which costs only
US$ 0.075, proved comparable to the Steiner trap which was the best.
Therefore, a screen of 80 such traps was established in the larger plantation
and serviced weekly during January — September 1973. Results show that
the population abruptly increased in March, reached a peak in April and
thereafter gradually declined.

In sterilization studies, a Gammacell 220 was used to irradiate the pupae 2 days before emergence with dosages ranging from 1 to 13 krad. None of these doses had any substantial effect on % emergence and sex ratio. Pairs made up of an irradiated female with a normal male and vice versa were segregated and their longevity, fecundity and fertility were measured. At doses below 13 krad, longevity was normal. At 3 krad fecundity was normal when the male only was treated but drastically reduced when the female was irradiated. Females treated with 5 - 13 krad did not lay eggs whereas fecundity was normal when the males were irradiated with doses below 13 krad (at this dose fecundity dropped concomittantly with longevity). Both pair combinations were completely sterile at doses from 5 to 13 krad.

STUDIES ON THE MEXICAN FRUIT FLY Anastrepha ludens LOEW

D. Enkerlin S
Instituto Tecnológico y de Estudios Superiores de Monterrey,
Monterrey, N.L., Mexico

The following gives the salient results of the work carried out on the Mexican fruit fly from 1971 to 1973.

In attempts to determine the optimum technique for immobilizing A. ludens adults, a cold treatment (6°C ± 5) and exposures to CO_2 and N_2 were compared for their effects on the time required for recuperation, fecundity, fertility and longevity. Adults were exposed to these treatments for 15, 20, 25 and 30 min: none of the treatments proved ideal, CO_2 having a more detrimental effect than the others.

In irradiation studies, when males from pupae treated with gamma rays were crossed with normal females, egg sterility increased from 37% at 0.5 krad to 100% at 6.0 and 8.0 krad; females treated with 2 krad or more did not lay eggs. Flies treated with 6 krad are being released in a suppression programme in Baja California Norte on the Mexican-US border.

Bioclimatic cabinets were used to study the development of all stages of the Mexican fruit fly under various regimes of temperature and light; RH was not controlled. The range for temperature was 18 - 28°C and for illumination 10 - 14 h. Eight different combinations were used including a stepwise decrease in both factors. It was found that larval development was fastest and pupal yield highest under a constant regime of 28°C and 14-h photoperiod. However, adult emergence was highest at 23°C and 12-h light throughout. Under all conditions studied, the sex ratio was somewhat in favour of the males.

Puparia were dissected on each successive day of the pupal development and examined morphologically to determine when the pupa is formed. No clear-cut difference could be found in between the successive stages inside the puparium, and development appears continuous; thus far, there is no evidence of a diapause.

Six compounds (terpinyl acetate, linalol, beta ionone, edeoma and extracts of Chapote amarillo and Schinus in McPhail traps were tested for their attractiveness to A. ludens adults released 90 - 100 m from the traps. The checks were another similar six traps baited with cotton seed hydrolysate. The hydrolysate proved a far superior attractant.

SEIBERSDORF CONTRIBUTIONS

Contributions from staff of the
Seibersdorf Laboratory, IAEA, Vienna

DECISION MAKING IN PEST CONTROL PROGRAMMES

B.A. Butt

Introduction

In planning a pest control programme, one must make a decision, using all available information, on what method or combination of methods of pest control will be used. This paper presents a systematic method[1] for reaching this decision by using a hypothetical situation and going through the steps for reaching a decision.

Situation

The golden fruit fly has been found in citrus near the airport in the city of Atoma. The citrus of Atoma has no market value, but the citrus-lined streets are a tourist attraction. The area is also famous for its unusual speciality fruits.

About 200 km west of Atoma is a 50×10^6 US dollars/year citrus industry. The pests attacking this citrus are held below economic levels by biological control and no pesticides are applied; as a result, the non-sprayed fruit is sold at a premium. If the golden fruit fly reaches this area, pesticides will have to be used, thus reducing the market value of the fruit. The pesticides will also cause outbreaks of secondary pests, and therefore additional applications of pesticides will be required. The fruit growers fear that the golden fruit fly could destroy their industry and therefore have raised US $\$5 \times 10^6$ to eliminate the pest in Atoma before it reaches the commercial citrus growing area.

Procedure

First the objectives must be defined. In this case the objectives are to eliminate the golden fruit fly in Atoma before it reaches the commercial citrus area. This must be done at a minimal cost and adverse affect on the environment.

In choosing a method of achieving these objectives, resource limitations and required results that must be met are listed. If a method does not meet these "must objectives" it is no longer considered. Areal sprays and host removal do not meet the "must objectives" (Table I) and therefore are no longer considered.

The objectives for the best use of resources and the maximum results with the minimal disadvantages are then listed. These "want objectives" are weighted 1 through 10, 10 being the most desirable. Each of the methods of control are then scored on a 1 to 10 scale for each of the "want objectives". These scores are then multiplied by the original weights assigned to each objective (Table I). The products are then totalled for each control method and compared. One can see that ground treatment and release of parasites were 266 and 259 respectively, whereas golden fruit fly attractant and the sterile-insect technique were 405 and 452 respectively.

---

[1] KEPNER, C.H., TREGOE, B.B., The Rational Manager, McGraw-Hill, New York (1965).

TABLE I.   EVALUATION OF VARIOUS CONTROL METHODS TO ELIMINATE THE GOLDEN FRUIT FLY FROM ATOMA

OBJECTIVES		Ground treatment with Malathion bait spray			Aerial spray of area with Malathion bait spray			Sterile-insect technique			Release of parasites			Host removal			Golden fruit fly attractant		
MUST		Information			Information			Information			Information			Information			Information		
Total programme cost must not exceed US $5 million		US $4.5 million	Go		US $3 million	Go		US $4 million	Go		US $4 million	Go		US $3.5 million	Go		US $3 million	Go	
Must not use persistent or highly toxic insecticides		Non-persistent	Go		Non-persistent	Go		No insecticide	Go		No insecticide	Go		No insecticide	Go		No insecticide	Go	
Must have at least 95% probability of success		99%	Go		99%	Go		99%	Go		99%	Go		99%	Go		97%	Go	
Programme must meet local environmental codes		OK	Go		Aerial sprays not allowed	No go		OK	Go		OK	Go		OK	Go		OK	Go	
Programme must be acceptable to city of Atoma		OK	Go					OK	Go		OK	Go		Not acceptable because citrus-lined streets important to tourist trade	No go		OK	Go	
WANT	**Wt**	Information	S	WT x S	Information	S	WT x S	Information	S	WT x S	Information	S	WT x S	Information	S	WT x S	Information	S	WT x S
High degree of certainty that elimination will be achieved	10	99% based on aerial spray data	9	90				99% based on two related programmes	10	100	99% based on limited field tests	2	20				97% based on experimental plots	8	80
Minimal disturbance to residents of Atoma	5	Will cause some disturbance	3	15				Minimal	10	50	Minimal	10	50				Some in placing traps	8	40
Completion in the shortest possible time	8	3 years	5	40				18 months	10	80	5 years	2	16				2 years	9	72
Eradication at minimal cost	6	US $4 million	6	36				US $4 million	7	42	US $4 million	7	42				US $3 million	10	60
Minimal import of supplies and materials	4	US $3 million Malathion	2	8				None — flies reared on local material	10	40	None — flies reared on local material	10	40				US $1 million attractant	5	20
Programme acceptable to Provincial and National Depts of Agriculture	7	Acceptable	9	63				1st choice because of successes in two other cities	10	70	Question effectiveness	3	21				Agreeable	9	63
Should not cause secondary pest problems	7	Problems with scales and mites may develop	2	14				No problem	10	70	No problem	10	70				No problem	10	70
TOTAL			266						452			259						405	

ALTERNATIVES

Wt   = weight 1-10
S   = score 1-10
WT x S = weighted score

TABLE II. EVALUATION OF ADVERSE CONSEQUENCES OF VARIOUS CONTROL METHODS TO ELIMINATE THE GOLDEN FRUIT FLY IN ATOMA

Ground treatment with Malathion bait

1. Some residents may refuse to allow spray crews into gardens	Probability	8	
	Seriousness	10	
	P × S		80
2. Tourists complain of spray odour and spots on cars	Probability	4	
	Seriousness	2	
	P × S		8
	Total		88

Sterile-insect technique

Flies must be produced in non-host area	Probability	10	
	Seriousness	4	
	P × S		40
	Total		40

Release of parasites

Parasite not effective in cool weather which occurs in some years	Probability	3	
	Seriousness	10	
	P × S		30
	Total		30

Golden fruit fly attractant

1. Recent research shows that the golden fruit fly attractant is also highly attractive to the key pollenators of the speciality fruits produced in Atoma	Probability	10	
	Seriousness	10	
	P × S		100
2. The odour of the golden fruit fly attractant is offensive to some people	Probability	8	
	Seriousness	5	
	P × S		40
3. Tourists steal traps	Probability	8	
	Seriousness	3	
	P × S		24
	Total		164

Now the possible adverse consequences of each control method must be listed. The probability and seriousness of each of these are rated 1 through 10. The probability and seriousness scores are then multiplied and the sums of the products for each control method are compared (Table II). The figures developed in the adverse consequences cannot be subtracted from the performance total of the want objectives, but must be compared separately.

Decision

Two control methods, sterile-insect technique and golden fruit fly attractant, have very high performance ratings; however, the golden fruit fly attractant has very serious adverse consequences. Therefore, the sterile-insect technique should be used to eliminate the golden fruit fly from Atoma.

PRACTICAL PROBLEM ANALYSIS OF A STERILE FRUIT FLY RELEASE PROGRAMME

B.A. Butt

Any operational programme, whether simple or putting a man on the moon, requires detailed planning in advance. Problems are best solved by preventing them; however, potential problems must be identified and contingent plans developed. This paper presents a problem analysis of a typical sterile fruit fly release programme. An analysis of a specific programme would differ as would the probability and seriousness of a problem.

The reader will note that "trained personnel" and "monitoring" or "quality control" appear a number of times in the analysis. The need for well-informed and well-trained personnel as well as a detailed monitoring and quality control system cannot be over-emphasized.

POTENTIAL PROBLEM ANALYSIS OF A STERILE FRUIT FLY RELEASE PROGRAMME

Potential problem	Likely cause	Preventive action	Contingent action	Trigger contingent/ progress report	Probability/ seriousness
Rearing failure	Shortage of diet ingredients	1. Order well in advance 2. Maintain strict inventory 3. Provide proper storage facilities 4. Test all new shipments in advance 5. Do not store all supplies in one location	1. Have alternative food if normal ingredients are not available 2. Have list of suppliers other than the normal one 3. Maintain contact with other workers using same materials	1. Low inventory 2. Quality control shows ingredients unsatisfactory	3/9 = 27
	Disease	1. Sanitary rearing 2. Train personnel	Have separate colonies	Quality control	1/7 = 7
Low production	Adult, larval or egg mortality Equipment failure	1. Train personnel 2. Maintain full time workers 3. Buy good equipment for controls and monitoring 4. Bring production to full release level before flies are required in the field	1. Over production to correct	1. Monitoring of rearing 2. Monitoring of climatic conditions in rearing rooms	9/2 = 18
	Wrong amount of ingredient(s): too little or too much	Have a check list of amounts of ingredients	Prepare new lot of diet	Poor insect production	4/2 ≠ 8

Potential problem	Likely cause	Preventive action	Contingent action	Trigger contingent/progress report	Probability/seriousness
Insect mortality in transit or during release	1. Overheating or cooling 2. Shipping time too long 3. Release equipment damaging to insects 4. Exposure to harsh climatic conditions in field 5. Predators	1. Maintain good shipping temperatures 2. Have direct shipments of minimal time or do not rely on public transportation 3. Ship in several containers 4. Safe release equipment 5. Release in cool part of day 6. Release when and where predators are minimal; may need protective release stations	1. Have alternative method of transportation and releases 2. Have alternative release schedules and routes	1. Monitoring of insects at release sites 2. Poor recovery of released insects in traps	5/8 = 40
Release failure	Aircraft or vehicle failure	Good maintenance	Standby pilot or driver. Alternative release method	Report of failure	1/3 = 3
	Pilot or driver sick or on leave	Plan leave	Standby pilot or driver. Alternative release method	Report of absence	1/3 = 3
	Bad weather	Have ground release stations	Alternative release method	Weather report	3/1 = 3
	Failure of release equipment	Good maintenance and check up	Alternative method of release. Standby equipment	Report of equipment failure	2/2 = 4
	Not allowed to fly over or enter property	Good public relations and take care not to fly over livestock and homes at low level	Alternative method of release. Heavy releases surrounding prohibited area	Complaint	1/2 = 2

POTENTIAL PROBLEM ANALYSIS OF A STERILE FRUIT FLY RELEASE PROGRAMME (cont.)

Potential problem	Likely cause	Preventive action	Contingent action	Trigger contingent/ progress report	Probability/ seriousness
No overflooding by flies	Not enough flies released	1. Over produce in rearing 2. Spray sources of high infestation	Spray area	Trap monitoring infested fruit	3/8 = 24
	Fly-in of native flies	1. Over produce in rearing 2. Spray sources of flies	Spray sources of flies	Trap monitoring infested fruit	4/5 = 20
	Mortality of released flies	1. Handle with care 2. Train personnel	Correct handling	Laboratory and field monitoring	4/6 = 24
	Released flies not competitive	1. Continual monitoring of reared flies 2. Maintain more than one colony	Use alternative colony	Laboratory and field monitoring	4/6 = 24
Release of fertile flies	Flies were not irradiated	1. Have rigid protocol in irradiation procedures 2. Have fail-safe device on irradiation 3. Train personnel	1. Correct procedure 2. Spray release area if needed	1. Hatch samples in monitoring 2. Field infestation	1/10 = 10
	Flies received a low dose of irradiation	1. Same as above 2. Have double timer	1. Correct timer 2. Spray release area if needed	1. Hatch samples in monitoring 2. Field infestation	1/9 = 9

TAGGING OF ADULT MEDITERRANEAN FRUIT FLY,
Ceratitis capitata (WIEDEMANN), WITH RUBIDIUM

A. Wehrstein

When large numbers of marked insects are needed, e.g. in release/recapture studies, tagging with radioisotopes is inconvenient. Dyes may affect the performance of the marked insect. As an alternative to these techniques we tested detection by absorption spectroscopy of rubidium in adult medflies.

The marker was administered as rubidium chloride at various concentrations either in the standard larval medium or in the diet of the newly emerged adult. For detection of the marker, the adult flies were killed by cooling and ashed at 500°C for 5 h. The ashed sample was dissolved in 0.5 ml of 0.1$\underline{N}$ perchloric acid with little heating. One ml or more of acid was used to rinse the crucible. This solution was transferred to glass ampoules and then read directly with the atomic absorption instrument. Results are presented in Tables I - III.

Table I shows that on day 0 the rubidium concentration in the treated fly increased with the level of the marker in the diet; thereafter, the detectable amount of rubidium dropped rapidly and levelled off between days 11 and 22. The biological half-life was about 2 days.

Table II shows the average amounts of rubidium detected in larvae, pupae, puparia and teneral flies when the marker was administed in the larval medium.

Table III gives the average amounts of rubidium detected in 48-hour-old flies fed during that period various concentrations of the marker in 10% aqueous sucrose solution.

TABLE I. AVERAGE AMOUNTS[a] OF RUBIDIUM IN MEDFLY ADULTS REARED AS LARVAE ON DIETS CONTAINING VARIOUS CONCENTRATIONS OF THE MARKER

Day	Amount of rubidium in the diet in ppm			
	0	500	1000	2000
0	0.1083 ± 0.0248	7.3305 ± 0.6038	11.6422 ± 0.5186	20.9002 ± 2.9028
2	0.0759 ± 0.0016	2.2680 ± 0.8825	5.0780 ± 1.0837	10.1404 ± 4.0794
8	0.0728 ± 0.0217	0.7354 ± 0.2322	0.7663 ± 0.4180	3.3053 ± 3.2279
11	0.0449 ± 0.0147	0.2709 ± 0.0774	0.3716 ± 0.1858	0.4412 ± 0.1703
18	0.0464 ± 0.0132	0.2322 ± 0.1161	0.2554 ± 0.1006	0.9831 ± 0.9599
22	0.0774 ± 0.0286	0.2632 ± 0.1084	0.5186 ± 0.1626	0.4954 ± 0.0697

[a] $(\bar{x} \pm S.D.)$ µg/adult medfly

TABLE II.   AVERAGE AMOUNTS[a] OF RUBIDIUM DETECTED IN
LARVAE, PUPAE, PUPARIA AND TENERAL FLIES WHEN MARKER IS
INCLUDED IN THE LARVAL MEDIUM AT VARIOUS CONCENTRATIONS

Organism	Amount of rubidium in the diet in ppm		
	500	1000	2000
Larvae (3rd instar)	11.1058 ± 0.4021	15.9089 ± 1.1506	19.8388 ± 1.2113
Fresh pupae	12.1207 ± 0.9691	17.3708 ± 3.2813	23.9982 ± 2.6075
Nine-day old pupae	10.7670 ± 0.5262	16.5704 ± 1.5568	23.6992 ± 3.7702
Puparia	2.3304 ± 0.1921	3.2641 ± 1.9737	5.7671 ± 0.7696
Teneral flies	7.3305 ± 0.6038	11.6422 ± 0.5186	20.9002 ± 2.9028

[a] $(\bar{x} \pm S.D.)$ µg/organism

TABLE III.   AVERAGE AMOUNTS[a] OF RUBIDIUM DETECTED IN TWO-
DAY-OLD ADULT MEDFLIES FED RUBIDIUM CHLORIDE DURING
FIRST 48 HOURS[b]

	Amount of rubidium in adult diet in ppm		
	500	1000	2000
Flies not washed	17.8 ± 1.9	26.3 ± 5.5	43.6 ± 8.8
Washed	17.6 ± 4.5	30.0 ± 5.3	47.8 ± 13.2

[a] $(\bar{x} \pm S.D.)$ µg/fly.
[b] in 10% aq. sucrose solution.

REARING OLIVE FLY AT SEIBERSDORF

Anne Tischlinger

    The following is a description of the methods and equipment used for
maintaining stock colonies of the olive fly.
    The adult flies are maintained in rotating cylindrical sleeve cages
(36 cm in diameter, 190 cm high) at the rate of 500 females and 100 males
per cage.   Each cage has six wax cones for oviposition.   The flies are fed
a 4 : 1 mixture of carbohydrate (sucrose: 9 parts; fructose: 1 part) and
yeast hydrolysate.   Water is offered separately through wicks.   At the end
of 6 weeks the cage is discarded.
    Eggs are collected daily, washed for 2 min in 1% formaldehyde, rinsed
with distilled water and held over distilled water in cups with nylon-cloth
bottoms until used.

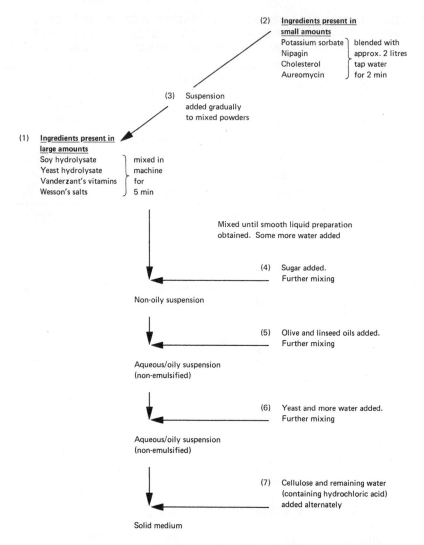

FIG. 1.   Preparation of olive fly larval medium.

The composition of the medium used for larval rearing is shown in Table I. The preparation sequence of this medium is indicated in Fig. 1. The ingredients are mixed in a cement mixer and then remixed in a house-hold mixer to obtain proper texture. The medium is dispensed into paper cups at the rate of 100 g/cup. Each cup is inoculated with about 435 eggs in a very small volume of distilled water and the cup is closed. The cups are held at 25°C and, after 8 days, small holes are punched in the bottoms to allow the mature larvae to leave the medium for pupation. Pupation takes place in bran placed in trays below the cups.

TABLE I. INGREDIENTS IN ONE KILOGRAM OF SOLID LARVAL DIET

Water	480	ml
Olive oil	80	ml
Potassium sorbate	0.5	g
Nipagin	1.0	g
Aureomycin	50.0	mg
Sugar	40	g
Linseed oil	3	ml
Brewer's yeast (dried)	100	g
Soy hydrolysate	30	g
Yeast hydrolysate	10	g
Vitamins (Vanderzant)	10	g
Cholesterol	1	g
Wesson's salts	5	g
1N HCl	adjust to pH	4.0
Cellulose	227.5	g

Larval development is completed in 10 - 16 days, the mortality being 70 - 75%. The individual pupal weight varies from 2.9 to 5.6 mg. Adults emerge after another 10 - 11 days at the rate of 20 - 70%.

ADULT OLIVE FLY DIETS: A COMPARISON

A. Wehrstein

The effects of six diets for adult olive flies on the fecundity, fertility as well as the male and female longevity of this species were compared. Five of these diets have been used by other workers, one was formulated in Seibersdorf. The composition of the diets and their origin is shown in Table I.

To ensure uniformity, the flies were fed these diets during the same period and under the same conditions. The flies were kept in small, round, plastic cages with a concentric wax cylinder for oviposition. Each cage contained 30 females and 5 males; there were 2 cages/diet. Food was presented on filter paper strips hanging from the ceiling of the cage; it was renewed daily. A sugar cube and distilled water were also present and renewed as required. The results are shown in Figs 1-6. According to these data, diet F was optimal as it resulted in a reasonable longevity combined with high fecundity and fertility.

TABLE I. DIETS USED BY DIFFERENT INVESTIGATORS FOR ADULT OLIVE FLIES

	A	B	C	D	E	F
	Seibersdorf	Moore (1959	Hagen et al. (1963)	Santas (1965)	Economopoulos and Tzanakakis (1967)	Orphanidis et al. (1969)
Water	10	50	50	50	50	10
Yeast hydrolysate (enzymatic)	10	10	10	10	10	
Partially hydrolysed protein[a]						10
Brewers yeast						30
Sucrose	36	40	40	20	40	
Fructose	4					
Honey				20		50
Egg yolk					7	
Choline chloride			0.05	0.05		0.05
Streptomycin sulphate			0.25			

[a] Probably yeast hydrolysate (enzymatic).

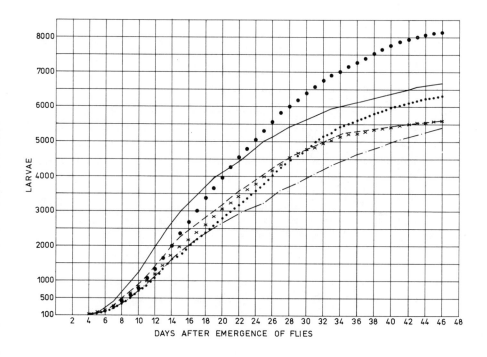

FIG. 1. Mean larval production per 30 females.
$--$ diet A
$....$ diet B
$-\cdot-$ diet C
x x x diet D
$-$ diet E
$\bullet\bullet\bullet$ diet F

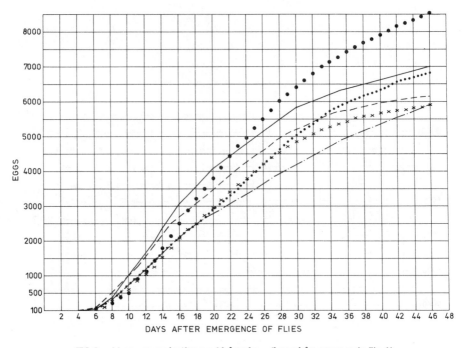

FIG. 2. Mean egg production per 30 females. (Legend for curves as in Fig. 1).

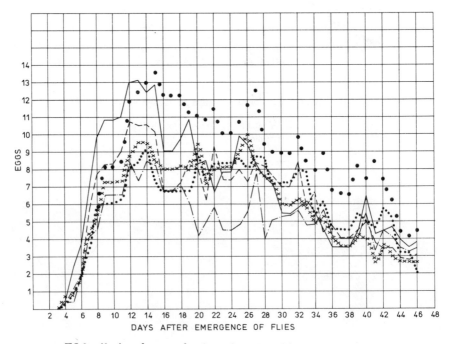

FIG. 3. Number of eggs per female per day. (Legend for curves as in Fig. 1).

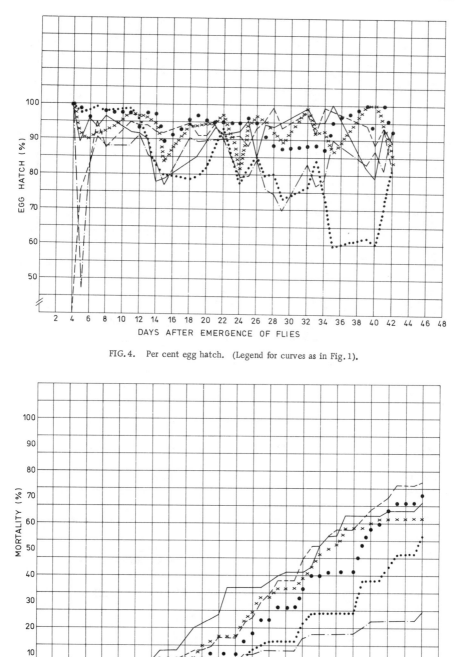

FIG. 4. Per cent egg hatch. (Legend for curves as in Fig. 1).

FIG. 5. Female mortality. (Legend for curves as in Fig. 1).

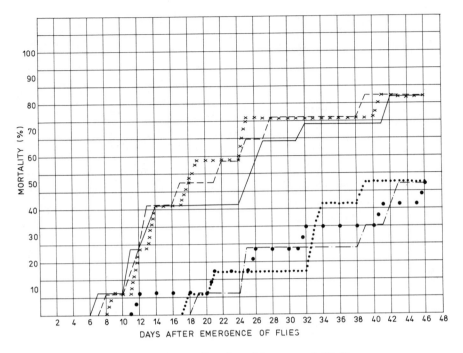

FIG. 6.   Male mortality. (Legend for curves as in Fig. 1).

REFERENCES

ECONOMOPOULOS, A. P. , TZANAKAKIS, M. G. , Egg yolk and olive juice as supplements to the yeast
hydrolysate-sucrose diet for adults of Dacus oleae, Life Sci. 6 (1967) 2409.

HAGEN, K. S. , SANTOS, L. , TSECOURAS, A. , "A technique of culturing the olive fly, Dacus oleae Gmel. ,
on synthetic media under xenic conditions", Radiation and Radioisotopes Applied to Insects of Agricultural
Importance (Proc. Symp. Athens, 1963), IAEA, Vienna (1963) 333.

MOORE, I. , A method for artificially culturing the olive fly (Dacus oleae Gmel.) under aseptic conditions,
Ktavim 9 (1959) 295.

ORPHANIDIS, P. S. , PETSIKON, N. , PETSEKOS, P. G. , Eléments sommaires concernant un élevage
expérimental du Dacus de l'olive sur substrat artificiel, In 8th FAO ad hoc Conference on the Control of Olive
Pests and Diseases, Athens, 8 - 12 May 1969.

SANTAS, L. A. , On a new synthetic nutritive substrate for rearing larvae of Dacus oleae Gmel. (Diptera-
Trypitidae), Bull. Agr. Bank Greece 144 (1965) 8.

LIQUID LARVAL DIET FOR THE OLIVE FLY Dacus oleae GMELIN

A. Mourad

Various absorbent materials were tested with the liquid diet developed by Mittler and Tsitsipis[1]. The absorbents were cotton wool, cotton and paper towelling. These were autoclaved at 120°C for 45 min and then put in 20 cm × 12 cm × 10 cm plastic boxes rinsed in 70% ethanol for asepsis. The bottom of each box was perforated with 50 2-mm-diameter holes. The holes were masked with tape until the larvae were ready to pupate. Pupation took place in bran provided below the boxes.

The liquid diet was poured onto the substrates in the boxes to give 3 - 6 g diet/g substrate. After infestation with eggs, the boxes were kept tightly shut in the dark at 25°C for 12 days. Under the conditions described, the development of fungi and other microorganisms was a problem. Results are shown in Table I.

TABLE I. EVALUATION OF THE MITTLER-TSITSIPIS LIQUID DIET ON VARIOUS SUBSTRATES

Substrate	No. of eggs planted	Eggs/g diet	Mean No. of pupae	Mean pupal weight	Mean % emergence
Cotton wool	500	20	105	5.3	48
	200	8	109	6.0	78
Cotton towel	500	25	87	5.1	51
	200	10	87	5.3	47
Paper towel	500	18	25	4.4	17
	200	10	12	4.2	10

---

[1] MITTLER, T.E., TSITSIPIS, J., 1973. Economical rearing of larvae of the olive fruit fly, Dacus oleae, on a liquid diet offered on cotton towelling, Entomol. Experim. Appl. 16 (1973) 292.

STATEMENTS AND RECOMMENDATIONS

GENERAL STATEMENTS AND RECOMMENDATIONS

1.   The Panel considers fruit flies a major economic pest problem, especially in developing countries. New important fruit production areas are being threatened because some Trypetids are currently expanding their geographic distribution, e.g. the Mediterranean fruit fly in Central America. Likewise, quarantine regulations often limit potential markets. Heavy reliance on insecticide applications has led to new problems in the management of agricultural pests, namely insecticide resistance, resurgence of secondary pests, undesirable chemical residues and environmental contamination. The sterile-insect technique (SIT) is one of the few highly specific control methods which can be used to overcome these problems. As shown here and in previous panels, the effectiveness of this technique has been demonstrated in quarantine, economic suppression and eradication, thereby stimulating research on target species which otherwise might not have been carried out.

2.   The prominent role of the Joint FAO/IAEA Division of Atomic Energy in Food and Agriculture in promoting the development and application of the SIT for the control of fruit flies is recognized and highly commended by the Panel. The emphasis in this work has been predominantly for the benefit of the developing countries but the impact also has been felt in more technologically advanced countries. Therefore, the Panel strongly recommends that the Joint Division continue and expand its programme of fruit fly suppression by the SIT.

3.   In view of the budgetary constraints limiting the assistance the Joint Division can render, the Panel strongly recommends the appropriation of monies for the establishment of a revolving fund. The sum required would be US $250 000 annually, to be administered by the Joint Division for the specific purpose of assisting member nations in the control of fruit flies by the SIT. Such a fund would permit a developing country with an ongoing programme, but lacking the funds for bringing it to completion, to obtain the necessary support.

4.   In the urgency to develop alternatives to pesticides, entomologists have tended to embark too rapidly on ad hoc release programmes frequently aimed directly and solely at eradication. Other possibilities such as the integration of the SIT with chemical methods and the use of the technique as a tool in pest management are only now receiving greater attention.
   A good example of the judicious integration of the SIT in a pest management scheme for eradication is the Peruvian medfly project. A recent UNDP study[1] supports the concept that eradication of the medfly in isolated valleys in Peru using the SIT in an integrated approach with insecticides is technically feasible and economically sound. The Panel fully supports the UNDP study team recommendations and urges that they be implemented.

---

[1] BRAZZEL, J.R., DAVIS, V.M. (1973), Peru Fruit Fly Eradication, Report of the Evaluation Mission, UNDP, 46 pp + 4 annexes.

5.   The Panel feels that the SIT is a tool to be used with discrimination. Its hasty application led and still leads to neglecting or awarding low priority to the collection and utilization of biological information. In this connection, the Panel considers ecological and behavioural information to be of particular importance. In addition, the Panel also strongly recommends to member nations and the Agencies involved to conduct economic analyses before commencing large-scale SIT practical control programmes.

6.   Whenever technical assistance is requested by member nations for large-scale fruit fly control by the SIT, economic analysis of the situation should be a sine qua non condition. In this connection the Panel noted that the OIRSA[2] member nations are showing active interest in the eradication of the medfly from Central America and that the economic analysis completed in 1972[3] confirmed the project's feasibility. The Panel wishes to point out, however, that this project would only be successful if the following conditions were met:

(1) availability of the basic biological information;
(2) proper planning on which to make sound decisions;
(3) utilization of suitable numbers of trained personnel for the needed programme operations;
(4) establishment of a quality control system;
(5) continuous operation of a parallel research programme, and
(6) adequate funding.[4]

7.   In addition to economic data, the Panel re-emphasizes the need for due consideration of alternative methods such as male annihilation, olfactory and visual attractants, inundative releases of parasites, etc. The Panel strongly recommends the preparation of a weighing scheme including the above criteria to arrive at a decision whether, in a given situation, the SIT should or should not be used.

8.   The Panel expressed concern that although the sterile-insect technique had proven to be an effective pest management tool, its use has largely been restricted to relatively few countries. Utilization of the technique is still mainly in the hands of researchers. Ultimately the responsibility for widespread field use lies with regulatory, control and extension officials. These officials are largely unaware of how the technique can be integrated into a nation's ongoing pest management programme. The Panel recommends that training offered through the FAO/UNEP project on Integrated Pest Control be expanded to include national and regional symposia to educate these officials. These symposia would be followed by training courses to prepare national field workers in the day-to-day technical programme requirements. Expert input into this training programme would originate with the Joint FAO/IAEA Division.

---

[2] Organismo Internacional de Sanidad Agropecuaria, including Costa Rica, El Salvador, Guatemala, Honduras, Mexico, Nicaragua, Panama.

[3] HENNING, R.G. et al. (1972), An Economic Survey of the Mediterranean Fruit Fly in Central America, USDA Field Report 21, 81 pp.

[4] At the time of going to press, this project took a new urgency as the medfly is reported close to Mexico; see also R.H. Rhode, Appendix I of these Proceedings.

9.   During the proceedings it became clear that certain research projects
in progress in specialized laboratories in various parts of the world are of
particular relevance to the efficient and intelligent application of the SIT
(e.g. research in allozymes, acoustical analysis and behavioural studies).
Most frequently this work is pursued in developed countries and it is
suggested that relevant member governments be strongly encouraged to
provide adequate support to those projects. It is recommended that those
specialized laboratories be available to support SIT projects in developing
countries (see also paragraphs 17 and 19).

10.   The Panel noted the present status of Research Contracts and Agreements on fruit flies and felt that a better balance is needed in their
distribution. Thus the Panel recommends that the number of Research
Agreements be increased in order to stimulate the already mentioned
(paragraph 9) supporting scientific activities carried out in technologically
advanced countries.

11.   Upon review of the recommendations of the FAO/IAEA Expert Panel
on the Practical Use of the Sterile-Male Technique for Insect Control
(Vienna, 13 - 17 November, 1972)[5], the present meeting endorses and
re-emphasizes recommendations for continued awareness regarding fitness,
competitiveness and prevention of genetic and physiological deterioration
of mass-produced fruit flies. Likewise, the present Panel recommends
that the Joint Division act as a centre for the development of methods to
evaluate artificial diets for efficient mass rearing, to evaluate mass
rearing techniques and to provide guidelines for the application of these
methods in developing countries. Thus, the following are recommended
as the primary objectives for the Joint Division's programme on fruit fly
suppression by the SIT:

    (a)   development of methods for the testing of media and other links
        in the fruit-fly production chain;
    (b)   improvement of irradiation methods;
    (c)   development of quality control techniques;
    (d)   improvement of logistics (devices and methods);
    (e)   training personnel.

12.   A primary requirement of the SIT is the designing of experiments for
mass rearing and to determine the quality of the insects produced both for
breeding stocks and for irradiation and release. Implementation of the
recommendations in paragraph 11 will require collaboration between two
research professionals: an insect nutritionist and a specialist in behavioural
physiology, both being experts in experimental design.
     The nutritionist will carry out research on the nutrition of target
species and develop therefrom experimental procedures for the evaluation
of diets. The result of this work will appear in publications and will be
used in the training of fellows from developing countries. This work will
be done in constant co-operation with the behavioural physiologist so that
a trainee will have the benefit of experience in nutrition, the formulation
of diets and the development and conduct of some simple behavioural tests
for evaluation of insect quality.

---

[5]   The Sterile-Insect Technique and its Field Applications, Panel Proceedings, IAEA, Vienna (1974).

A further task of the behaviourist will be to collate information on quality-control research carried out in other locations. At the termination of his employment, this professional's experience and the information collected will be published in a laboratory and field manual.

The Panel further recommends that project leaders collaborate by corroborating under their own field conditions some of the techniques developed at the Agency.

13. The Panel recognizes that a problem exists as regards placement of trainees from developing countries due to the limited number of centres with ongoing programmes of fruit fly control by the SIT.

It is recommended that the allocation of trainees from developing countries be not restricted to ongoing projects but that use be made of the large number of existing centres of formal instruction.

The Panel recommends that the present level of training activities at the Seibersdorf laboratory be maintained and if possible increased.

14. The Panel noted with concern that in some developing countries Agency publications in entomology are not passed down from the administrative to the working level. Therefore, the Panel recommends that the Agency explore ways to ensure dissemination of its publications in entomology (in particular Panel Proceedings) so that they reach the maximum possible number of researchers in developing countries.

SPECIFIC STATEMENTS AND RECOMMENDATIONS

Operational aspects of fruit fly control by the sterile-insect technique

15. Having considered the minimum size area over which practical fruit fly control by the SIT could be expected, the Panel concluded this could not be determined with any degree of precision because of the numerous variables involved. Assuming adequate quarantine control, manpower, fly production and release capability, a key factor is the degree of isolation. Other factors which may impose a limit in effectiveness are the level of control by bait sprays, clean cultivation, and the degree of freedom given to the project manager in the implementation of the programme.

16. The Panel also examined the question of sterile stings by released females, possible counteraction by releasing males only and the beneficial role, if any, of the sterile female. The Panel considers sterile stings a marketing problem. The importance of those stings varies with the fruit fly species, the type of fruit (it may be important in some stone fruit and citrus varieties) and the time of the year. If the stings are a problem in experimental studies, the fruit may have to be bought. However, the problem is temporary where eradication is the goal and can be diminished by such methods as stopping or reducing releases before susceptible fruits come into production, by suitable cultural practices and selective spraying of early ripening varieties. Currently no feasible method exists to separate the sexes in large numbers; research to find methods to do this economically (e.g. by use of sex distortion) should be encouraged. The evidence thus far available suggests that, in fruit fly species with polygamous behaviour, the release of sterile females contributes to suppression.

17. The Panel considers of paramount importance developmental research
in quality control and quality control techniques. This is essential for
determining the most important biological parameters for sterile fly
releases. Such techniques will allow the early screening of the flies best
suited for a field programme as well as the continuous monitoring of the
effectiveness of the released flies. Quality control techniques are needed
for each fruit fly species. Since there are laboratories specialized in
studies on quality control of mass-reared insects, the services of these
laboratories should be utilized in SIT programmes.

18. A measurement of the effectiveness of sterile flies in mating with
wild flies is essential. This may involve various aspects of mating
behaviour. Once known, these can be used to develop monitoring systems
in a sterile-insect release programme. More field studies are needed on
the mating behaviour and ecology of fruit flies. In this respect, male
flies with sperm marked by radiotracers may be released in field experiments to establish mating frequency and competitiveness under natural
conditions.
      Radiotracer sperm tagging is necessary for special applications, but
the Panel does not consider it essential for general use during a sterile
release programme.

19. The Panel recommends that the Agency assist and co-ordinate studies
of fruit fly behaviour and ecology as related to the SIT. Monitoring genetic
variability (e.g. by allozyme analysis) is essential as an alarm system to
warn of increasing homogeneity in the breeding stock; the Panel recommends
that this be done in a central location to which sample material could be sent.

20. Where a fruit fly pest management programme calls for a multidis-
ciplinary approach, including extension and regulatory practices, the Panel
recommends that the SIT be integrated with other suppression methods,
such as male confusion and male annihilation, the judicious use of bait
sprays, trapping, parasites and predators.

21. The Panel considers the airplane an essential tool in large-scale control
and eradication programmes and therefore strongly recommends that
provision be made to include this piece of equipment at the beginning of
such programmes. In this connection improvement and evaluation of aerial
release techniques are required. In smaller programmes, before deciding
on air releases, due consideration should be given to economic factors,
such as size of the area, cost of labour, environmental conditions and
topography.

22. A further need is some automation to reduce human error, increase
the quality of the product and lower costs. The Panel recommends that
resources be allocated to the development of such devices, particularly
in large-scale projects requiring around-the-clock work.

23. The Panel had before it a scheme designed to assist in the process
of decision making[6] in the conduct of an SIT programme. Such a scheme

------

[6] See B.A. Butt, these Proceedings.

is strongly recommended for the consideration of those contemplating
the application of the SIT.

Mediterranean fruit fly Ceratitis capitata (Wiedemann)

24. Information available from various medfly rearing units (Costa Rica,
Cyprus, Israel, Italy, Spain, Austria) allows an approximate calculation
of the space requirements for mass producing the medfly. The present
norm is approximately 50 000 to 60 000 insects per week per m^2 total rearing
facility surface (i.e. irradiation and storage rooms, office, etc. included).

25. Production of the medfly is still relatively expensive because of costly
nutrients such as protein hydrolysate and brewer's yeast. In most cases
these ingredients are imported and in developing countries this often is a
drain on hard currency. Therefore, the Panel strongly supports the current
Joint Division studies on the replacement of expensive nutrients by inexpensive and readily available materials. The Panel also recommends that the
Joint Division encourage similar work in developing nations.

26. Regarding winter releases of sterile medflies to protect citrus, the
Panel considers this a good approach if the climate does not inactivate the
insects, there is mating, the population is low, and the market value of the
fruit is not reduced by sterile stings.

27. The Panel recognizes that the effectiveness of the SIT for medfly control
during spring and summer has been repeatedly demonstrated, e.g. in Spain,
Italy and Israel. Good results also were reported after combining this
technique in pest management with chemical and biological control agents,
e.g. in Argentina and Peru. However, to convince administrators to put
this technique into practical use, a demonstration of eradication could be
decisive. Therefore, the Panel urges the FAO/IAEA to co-ordinate a
campaign of medfly eradication in a Mediterranean country drawing upon
the expertise and pooling the resources of several countries in the area.
The assistance of the International Organization of Biological Control in
such a venture should be enlisted (see Appendix II).

Olive fly, Dacus oleae (Gmelin)

28. The biotechnology for olive fly production is steadily improving.
According to information from Greece the present olive fly production norm
is approximately 5 000 pupae per week per m^2 total rearing facility
surface.

29. The present cost of olive fly production in Greece is US $1/1 000 pupae,
labour included. The Panel is not convinced that cost should be a primary
consideration in substituting imported nutrients for locally available ones
because imported products are more standardized and therefore
their use in the various rearing facilities would eliminate a source of
variation and increase the validity of comparisons. Nevertheless, the
Panel recommends that the Joint Division obtain a list of materials used
in olive fly rearing.

30. The Panel recognizes that the SIT can and should fit into a pest management system for olive fly control. Because of a lack of geographical isolation, eradication usually will not be the objective of the SIT in olive fly suppression programmes. The pest management scheme should comprise two sequential phases:

(1) Suppression: by cultural practices (spring collection of remaining fruits), treatment with selective insecticides (early summer), trapping (pheromones included);

(2) Control: by release of sterile olive flies in combination with parasites.

31. The co-operative Yugoslav/Joint Division/USDA olive fly research programme, including the development of better irradiation techniques, mass rearing, marking, trapping, release methods, diets, etc., complements research in other olive growing countries. The Panel notes that the Agency supplies olive flies of Greek origin to Yugoslavia for ecological studies; it suggests that releases of a Yugoslav strain would be preferable.

32. The Panel recommends increasing the number of contracts, and if possible, the monetary value of contracts, for research on olive fly rather than on medfly.

33. Training in olive fly research should take place in olive-zone laboratories wherever this research is sufficiently advanced.

34. The Panel recommends that the Joint Division, together with organizations such as the International Biological Programme, the International Organization of Biological Control and the European Economic Community, encourage and help organize annual or biennial meetings of olive fly workers to exchange information. These meetings should rotate between major olive fly research centres (Italy, Greece, Yugoslavia, Spain, Portugal).

35. The Panel recommends an exchange of visits between Agency personnel engaged in developmental research on olive fly and their colleagues in the olive zone to improve co-ordination.

Other fruit fly species

36. In the developing countries some of the most important fruit fly species are:

Anastrepha ludens: Mexican fruit fly;
A. fraterculus: South American fruit fly;
A. suspensa: Caribbean fruit fly;
Ceratitis rosea: Natal fruit fly;
Dacus dorsalis: Oriental fruit fly;
D. cucurbitae: Melon fly;
D. zonatus: Mango fruit fly;
Rhagoletis cerasi: Cherry fruit fly.

37. In the developed countries, several species of <u>Hylemya</u> are of economic importance to vegetable crops and, based on the experience obtained in Belgium and Holland, could be considered in the near future as target species for the SIT.

38. The Panel feels that the developmental work in several research centres on the species listed is adequate and need not be undertaken by the Joint Division. However, the Joint Division could greatly contribute in the <u>R. cerasi</u> programme by co-ordinating the work of European researchers.

39. The Panel considers that the use of local materials after their extensive evaluation may be the only economical and practical solution to mass rearing the other fruit fly species just mentioned. It is recommended that the Agency assist and/or co-ordinate the evaluation of such materials.

40. Likewise, as regards some of the species listed, the Panel recommends that the Agency encourage co-ordination among different workers on specific topics such as radiobiology, rearing, ecology, release methods, etc.

41. It is recommended that the Agency continue to train scientists on the fruit flies listed here.

Anastrepha <u>ludens</u>

42. The 1969 Panel on the Sterile-Male Technique for Control of Fruit Flies[7] recommended that an eradication experiment be conducted in an isolated area of about 1000 km². Steps have been taken in this direction but on a somewhat smaller scale. However, since the exact distribution, ecology and host relations of this pest in Mexico and Central America are not known, work on these aspects should be encouraged. Likewise, the application of the SIT in conjunction with other control measures should be more extensively considered.

Anastrepha <u>fraterculus</u>

43. Research on this species is sufficiently advanced to make a study of the effectiveness and ecological implications of simultaneously releasing <u>A. fraterculus</u> and <u>Ceratitis capitata</u> worthwhile. This would be a unique opportunity for testing the possibility of simultaneously suppressing or controlling several species by the same technique and exploring possible interactions of such a suppression.

Anastrepha <u>suspensa</u>

44. The study of this species has greatly progressed in the USA. Other countries in which this fly is a problem should be encouraged to take maximum advantage of these studies.

As they become of economic importance, other <u>Anastrepha</u> species might be considered for inclusion in the Agency's programme upon requests from the affected countries.

---

[7] Panel Proceedings Series, STI/PUB/276, IAEA, Vienna (1970).

Dacus dorsalis, D. cucurbitae, D. tryoni and D. zonatus

45. Methods of rearing, trapping and irradiation as well as sterility data and chemical control measures are available for these four species of Dacus. The feasibility of suppressing D. dorsalis, D. cucurbitae and D. tryoni by the SIT has been demonstrated. Efforts should now be made to integrate the SIT in a sound pest management programme in large areas. Wherever and whenever possible, eradication should be attempted.

D. dorsalis and D. cucurbitae have several hosts in common, even though they may not complete their life cycle in some. Again, there is an opportunity for studies of simultaneous suppression.

Release to establish "guardian population" barriers to prevent infestation or reinfestation of fly-free areas should be undertaken.

Since much work has been done on D. tryoni, contact with workers on this insect should be maintained.

Rhagoletis cerasi

46. This species is of great economic importance in certain countries and is widespread, attacking a high percentage of fruit if no control measures are applied.

Research on R. cerasi now shows that in the near future the eradication of this pest can be proven at least in some rather isolated areas.

Suppression of this pest by the SIT should be studied as part of a pest management project.

Especially in the case of this insect, the Panel strongly recommends that the Agency further personal contacts between all workers concerned.

Hylemya spp.

47. The feasibility of suppressing H. antiqua by the SIT has been demonstrated in the Netherlands. The Panel recommends that the Agency encourage further basic and applied research in developed countries on this and other Hylemya species, such as H. brassicae. Large-scale field trials should be initiated as soon as mass rearing and ecological studies have developed to a suitable level.

A MEDFLY ERADICATION PROPOSAL
FOR CENTRAL AMERICA*

R. H. RHODE
Citrus Insects Investigations,
USDA, ARS,
Weslaco, Texas,
United States of America

Abstract

A MEDFLY ERADICATION PROPOSAL FOR CENTRAL AMERICA.

A survey sponsored by US/AID revealed that in 1970 direct crop losses due to Mediterranean fruit fly Ceratitis capitata (Wiedemann) infestations throughout 10 774 km² (2.7 million acres) in Costa Rica, Nicaragua and Panama approximated US $2.4 million. Potential annual losses would reach US $6.8 million if the medfly to spread throughout presently uninfested Central America. The survey team members estimated that a 6-year eradication programme involving sterile flies, malathion mixed with protein bait, or malathion alone would cost US $30.8, US $25.7 or US $21.8 million, respectively. The subsequent revision of these estimates by agricultural officials of the OIRSA member countries resulted in a 5-year programme involving both malathion and sterile medflies at a cost of US $20.5 million. The financing would be provided by contributions from the governments of the USA, the five Central American countries, Panama, Mexico and the United Kingdom that would be deposited in CABEI to cover the operational costs. This sum would be repaid to CABEI within 10 years and would constitute the nucleus of an emergency fund to combat plant and animal diseases or pests within the OIRSA region. Additional monetary and/or technical support would be requested of UNDP, IAEA, the University of California and the Interamerican Institute of Agricultural Sciences. Intensive trapping for medflies within OIRSA countries that are ostensibly free of this pest and in Belice should be a prerequisite to an active eradication programme. Supplementary information concerning medfly ecology and methods of marking and aerial release is needed.

One of the world's most notorious fruit pests, the Mediterranean fruit fly, Ceratitis capitata (Wiedemann), made its first appearance in Central America near San José, Costa Rica, in 1955. Initial attempts to eradicate the medfly from the locally infested areas with insecticides were abandoned after the insect became generally established throughout the coffee-producing regions of the central highland plateau. Instead, the newly-created Organismo Internacional Regional de Sanidad Agropecuaria (OIRSA) presently undertook the responsibility of preventing the spread of the medfly to other Central American countries. OIRSA, an international regional organization funded by annual contributions from its member countries, is concerned with the control or eradication of plant and animal pests and diseases that threaten

---

* Liberal references have been made to Field Report 21, Economic Survey of the Mediterranean Fruit Fly in Central America, US Dept. of Agriculture, in co-operation with US Agency for International Development, 1972, and Informe III Reunion Extraordinaria de Jefes de Sanidad y Cuarentena Vegetal, 15-18 de Mayo de 1973, Organismo Internacional Regional de Sanidad Agropecuaria.

the agricultural economy of the five Central American countries, Mexico and Panama. Nevertheless, despite regulatory and quarantine measures instituted by OIRSA to halt the medfly advance, the pest was detected in the Carazo region of Central Nicaragua in 1961 and in Western Panama in 1963.

In an effort to develop non-insecticidal methods of reducing native medfly populations, OIRSA, with assistance provided by the United States and the Interamerican Institute of Agricultural Sciences (IIAS), initiated a laboratory-rearing programme in 1958 for the production of medfly parasites and, concurrently, began investigations of medfly sterilization with gamma irradiation.

Also, in 1964, a pilot test involving the release of 1 million sterile medflies per week over a 1-square mile area on the peninsula of Puntarenas, Costa Rica, was carried out under a joint US/AID-OIRSA agreement. The results were sufficiently encouraging so that OIRSA, with United Nation's Development Programme (UNDP) support, entered into a co-operative programme in 1965 with the International Atomic Energy Agency (IAEA) to determine the feasibility of eradicating the medfly from Central America by means of the sterile insect release method. In an experiment conducted from September 1968 to May 1969, approximately 40 million sterile medflies were released by air each week over a 48-km^2 coffee and citrus area in Nicaragua. Results of this test demonstrated that average deposition of viable medfly eggs, larval infestations in coffee fruits and recoveries of pupae from mandarin oranges were at least 90.0% less in the release area than in the two checks [1].

At the conclusion of the joint UNDP-OIRSA-IAEA co-operative programme, a conference was held in 1970 at San José, Costa Rica, to assess the results of the sterile fly release experiment in Nicaragua and to discuss a course of further action to rid Central America of this pest. In addition to officials of the foregoing organizations, other delegates represented US/AID, USDA, US Department of State, US Atomic Energy Commission, US Air Force and the Ministries of Agriculture of each of the five Central American countries, Mexico and Panama. It was proposed and agreed that before considering a medfly eradication programme, US/AID would sponsor a study to determine (1) the economic losses sustained by the countries presently infested with the medfly, (2) the potential damages if the fly were to spread into previously non-infested countries and (3) to estimate the cost of eradicating the medfly from Costa Rica, Nicaragua and Panama.

Consequently, during 1971-72, a four-man team of specialists in entomology, tropical horticulture, agricultural marketing, and agricultural economics made an assessment of the medfly situation in Central America. The team at that time considered that self-sustaining medfly populations were present throughout 1 672 km^2 in Nicaragua, 7 000 km^2 in Costa Rica, and 2 102 km^2 in Panama, representing a total area of 10 774 km^2 (Fig. 1).

Medfly infestations in these countries are generally found in developed highland areas cultivated to coffee and citrus and in small towns and settlements along the Pacific coastal plain where tropical almond (Terminalia catappa) and mixed fruit plantings occur.

Citrus is the only commercial host fruit crop — except coffee — grown on an economic scale. Coffee is a favoured host fruit of the fly and, even though this crop sustains the major medfly infestations in Central America, losses in coffee bean production were estimated to be only about 1%. Of the

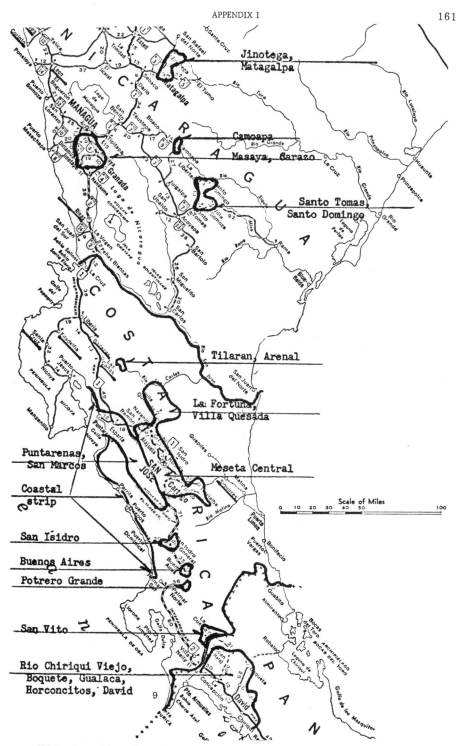

FIG. 1. Areas with a self-sustaining population of Mediterranean fruit fly (<u>Ceratitis</u> <u>capitata</u>).

citrus varieties, sweet orange and mandarines are the most heavily attacked. Medfly losses to fruit production in the infested Central American countries were estimated to be about 28% for sweet oranges, 50% or more for mandarines, 24% for grapefruit and 2% for other susceptible fruits.

Applying these estimates of susceptible crop losses to the estimated value of fruit production in the current medfly area, the value of losses due to this insect in 1970 was about US $2.4 million. Should the medfly spread to all areas of Central America, the same loss rates would cause a potential annual damage of US $6.8 million. Were the medfly to invade Belice and Mexico, the citrus industries of these countries would suffer annual losses of US $1 million and US $5 million, respectively. Potential damages to the US fruit industry would be much greater. At one-half the damage rates observed in Central America, medfly losses to the US citrus industry would approximate US $85 million annually. Because of the susceptibility of deciduous fruits to medfly attack, losses could amount to US $200 million annually, a sum which represents a third of the total value of production.

In their report, the medfly team emphasized the importance of establishing and enforcing strict quarantine measures to prevent the medfly from being carried into areas where it is not established or where eradication is still incomplete. Internal quarantine operations within each country would entail the efficient control of terrestrial, aerial and maritime traffic at designated check points and at international boundary crossings. Even though a medfly eradication attempt is not undertaken in Central America, a quarantine programme would still be necessary to retard the advance of the fly into new areas. Furthermore, upon completion of the Pan American highway, linking Panama with Colombia, vehicular traffic will provide the possibility for more rapid movement of the medfly and other exotic pests and diseases of agricultural importance between South and Central America. The team recommended that quarantine stations be established in northern Nicaragua on the West Coast and Pan American highways to control traffic moving into Honduras and El Salvador. It was estimated that operational costs of the two new stations would amount to about US $269 000 for a 4-year period.

A thorough and continuous trapping programme throughout Central America was also considered a necessary prerequisite to any programme or activity to control or eradicate the medfly. In the opinion of the team, a survey trapping programme should be conducted whether or not eradication of the medfly is attempted. Such an effort, utilizing triangular carton traps with Trimedlure incorporated in the adhesive material, would require external financing in the amount of about US $2.4 million. It is advisable that the survey programme be in operation at least 1 year before any control or eradication attempt.

The economic survey team judged that because of rugged terrain and associated adverse weather conditions, 3 740 km^2 of the total infested area of 10 774 km^2 could not be adequately treated by conventional airplanes — neither with insecticides nor sterile flies — and that the ground application of malathion bait sprays would be needed. Because of the broad scope of a programme to eradicate the medfly from Central America, a 6-year multistage plan of operations was proposed which provides for 1 year of preparation devoted to advance trapping, construction, procurement of supplies, etc., followed by 4 years of major eradication effort during which one-fourth of the total infested area would be treated each year and, finally, 1 year of final "mop-up".

Three alternate methods of eradicating the medfly were proposed: the use of malathion alone applied by air at 2 fl oz/acre, malathion at 2 fl oz/acre mixed with protein bait at a 1:4 ratio and the release of sterile flies. Although ultra-low-volume (ULV) malathion applications alone have not yet been proven completely effective against the medfly, it is known to cause a quick knock-down of flies [2].

The C-123 multiengined jet-assisted airplane was considered the most practical type for large-scale spray operations. Cities and towns lying within the air-treatable areas would receive coverages of malathion alone. Chemical treatments would consist of eight weekly spray applications for each of the insecticide programmes and would commence at the end of the wet period in November.

Marked medflies sterilized with 9 krad of gamma radiation would be released from DC-3 airplanes at the rate of 400 000 per km^2 per week. A laboratory-rearing facility capable of producing 500 million flies per week would be centrally located at San José, Costa Rica. Sterile-fly releases would begin during the period of low wild-fly population densities in September.

Because of potentially bad weather conditions and areas that may be missed in the application of the alternate methods of medfly eradication, some retreatment would undoubtedly be required. These costs were estimated at 5% of the total treatment cost for the current year of treatment and 10% for the following year for each type of eradication programme.

An intensive medfly-monitoring programme which would provide accurate and up-to-date information on sterile and wild fly density and distribution in the work areas is an essential element in an eradication project. It is estimated that the yearly cost of monitoring 10 sticky traps per km^2 per week over 1 000 km^2 would be about US $1.2 million.

Total costs of each of the proposed eradication techniques, used alone, including costs of retreatment and a monitoring programme, were estimated by the economic survey team at US $21.8 million using ULV malathion, US $25.7 million using malathion-bait spray, and US $30.8 million employing sterile fly releases.

In early 1973, OIRSA became increasingly alarmed with the consequences attendant upon the evacuation and displacement of a large portion of the population of Managua, Nicaragua, ravaged by earthquake in December 1972. To resettle in outlying areas, many of the homeless circumvented the internal medfly quarantine station situated on the Pan American highway, 22 km north of Managua, by dispersing along secondary roads and trails. Sporadic out-breaks of the medfly in northern towns previously freed of this pest indicated the probability that infested fruits from the major citrus producing region in central Nicaragua were being received in these areas to meet the demands of increased local consumption.

At an extraordinary reunion of the Ministers of Agriculture of the OIRSA member countries held in February 1973, OIRSA urged the immediate reinforcing of existing quarantine, trapping and control programmes to curtail the northward spread of the medfly. Considering the vulnerability of El Salvador and Honduras to invasion because of their proximity to medfly infestations in Nicaragua, OIRSA recommended a study be made of the AID Economic Survey Report to develop a technical-economic strategy to rid the

currently infested countries of this pest. Subsequently, in May 1973, officials
of Vegetable Health and Quarantine Departments from the OIRSA countries
met with members of OIRSA, US/AID (ROCAP), USDA, the economic survey
team and the Central American Bank for Economic Integration (CABEI) to
discuss strategy, costs and a possible manner of financing an eradication
programme.

From this conference evolved an integrated programme of up to 5 years
duration which would utilize insecticides and sterile insects to achieve
eradication through the following sequential approach:

(1) 6 ULV-malathion sprays at 1-week intervals
(2) 2 malathion-protein bait sprays at 2-week intervals
(3) 3 months of weekly sterile-fly releases at a rate calculated to
     maintain a minimal sterile-to-wild-fly ratio of 100:1.

The delegates considered the ground application of insecticides to the
area of 3 740 km^2 proposed by the medfly survey team to be economically
untenable and projected the use of malathion-bait sprays applied by heli-
copter as a means of reducing the estimated costs. Other revisions of
item costs presented in the Economic Survey Report resulted in an eradication
programme with an overall cost of about US $20.5 million composed of the
following operational expenses:

(1) Survey, trapping and quarantine .... US $1 859 826
(2) Supervision 2 029 157
(3) Insecticide applications 7 989 442
(4) Sterile fly production and release 6 734 200
(5) Administration 1 861 263

                              Total   US $20 473 888

As some possible means of obtaining the necessary funding for an
eradication project advanced at the May 1973 conference, it was proposed
that each OIRSA member country increase its own normal national quarantine
service support by 100% for the first year of the programme and by an additional
50% for the second year. Individual country support would continue at
this level until 2 years after the medfly is declared eradicated in Central
America. In addition, each OIRSA country would contribute a determined
amount to a fund established in CABEI for financing the initial phase of the
eradication programme and all countries would become current in payment
of their annual quotas. A grant contribution would be requested of the
Mexican Government to be used for operating costs. The amount of this
loan would be deposited in CABEI.

A loan would be requested of the United States Government through
US/AID (ROCAP) to be deposited in CABEI as part of a "Medfly Eradication
Fund" used to finance equipment, insecticides, local salaries, and other
costs. This fund would be administered by CABEI under the technical
direction of an OIRSA-USDA-IAEA-IIAS Technical Advisory Committee
to be established to guide this joint effort.

Portions of the US contribution would be made available to the Central
American countries for graduate training of national technicians in entomology
and short-course training of engineer agronomists and lower technician

levels in methods of trapping, quarantine, spraying, insect identification, etc.
Also to be paid from the US loan are the salaries and associated expenses for
6 USDA advisors for a 5-year period to assist in various aspects of the
programme as well as the costs associated with aerial spraying carried
out by the US Air Force in the event that assistance in the form of up to three
C-123 aircraft with ULV-dispensing equipment becomes available.

The UNDP would be requested to provide monetary assistance for an
expanded rearing facility and research and technical assistance support
through IAEA.

Other research assistance would be requested of US/AID-Technical
Assistance Board/University of California Pest Management Research
Project to work with IAEA and IIAS on insect radiation and release, research
related to aerial and ground spray systems, collecting and processing of trap
data and the provision of technical training programmes required by field
and other support personnel.

A request would be made of IIAS to provide research assistance on a
study to determine the effects of the medfly on tropical crops and to make
an analysis of the benefits to be derived by the Central American countries
through crop diversification and, also, to help in formulating a strategy including
location, crops, costs, etc. for the development of new crop production with
special emphasis on small and medium farm operations that could be implemented as soon as possible after the medfly becomes eliminated in Nicaragua,
Costa Rica, and Panama.

The British Foreign Assistance Agency would be requested to augment
its financial support to Belice for increased quarantine, trapping and related
programmes to prevent medfly introduction. In addition, the British
Government would be asked for a contribution to be deposited with other
funds in CABEI for costs associated with the eradication programme in
Central America.

Monies deposited in CABEI for use in defraying the operational expenses
of the eradication programme would be repaid to CABEI by the OIRSA member
countries in annual quotas within a 10-year period and would serve as the
nucleus of a revolving "Emergency Pest Eradication Loan Fund". This fund
would be available under conditions determined by the OIRSA directors for
the control or eradication of pests or for other problems considered emergency
situations by OIRSA.

The actual dollar values of contributions requested of the various
governments and international agencies will be considered following approval
of the financing plans by governments of the OIRSA member countries.

Before commencing an active eradication attempt in Central America,
it would seem advisable that intensive trapping surveys be conducted
throughout the OIRSA region and Belice to determine if the medfly has
encroached upon territories previously considered free of this insect.
It is not improbable that the scope of the proposed eradication effort would
be influenced by the number and size of any infestations discovered in the
countries north of Nicaragua. More detailed information is also needed
concerning medfly distribution within each of the currently infested countries,
particularly along the Atlantic littorals and in mountainous areas of restricted
accessibility. Remote-sensing techniques employing colour infra-red
aerial photography would provide a rapid and relatively inexpensive means
of detecting susceptible host fruit plantings or, alternatively, of eliminating the

need for extensive ground reconnaissance of undeveloped or agriculturally unsuitable regions.

Additional knowledge of wild-medfly life cycles, particularly the length of pupal development in different climatic environments, and the correlation of trap catches to native fly population densities, would be beneficial. Insect marking is a crucial element in sterile-insect release programmes. Existing methods for marking the adult insect, using fluorescent powders applied to the pupae, must be improved upon or practical alternative means found which would ensure virtually 100% identification of captured laboratory-reared medflies. It would also be desirable beforehand to compare methods of releasing sterile flies by air under conditions existing in the proposed work environment to determine whether medflies released in containers or discharged unconfined into the air stream constitutes the most effective and practical method. Electronic guidance systems to direct airplanes accurately along flight lines over forested areas must also be evaluated in advance under flying conditions present in Central America.

Even though the technological basis for suppressing wild medfly populations with chemicals and sterile insects has been established, the uneven terrain and variable weather patterns of Central America, in association with the general lack of communications, inadequate roads and landing fields in outlying areas, could present formidable barriers to an eradication effort.

Problems with logistics could be expected because of the time required to import supplies and equipment from markets outside the Central American region and the administrative delays involved in transporting material across international boundaries.

A tentative eradication strategy, advanced at the May 1973 conference, would exert a closing action on the medfly-infested region of Central America by the initial and simultaneous treatment of Nicaragua and Panama. However, in addition to the increased operational burdens created by a divided campaign, any eradication attempt in Panama would involve areas of southern Costa Rica because of medfly infestations in the border regions. Among other factors, a successful eradication enterprise would include the precise co-ordination of various specialized activities. It is unlikely that an untested task force would have the proficiency needed to overcome the obstacles imposed by widely separated and difficult work environments located in three different countries. OIRSA, therefore, may wish to consider concentrating the initial effort in Nicaragua rather than risk the possibility of the newborn programme becoming overwhelmed at the outset by a multitude of unresolved problems. Because of the nature of its widely dispersed areas with medfly infestations, Nicaragua, in itself, would provide an uncompromising proving ground, under working conditions similar to those which would be faced with confidence and expertise at a later time in Costa Rica and Panama.

REFERENCES

[1] RHODE, R.H., SIMON, J., PERDOMO, A., GUTIERREZ, J., DOWLING, C.F. Jr., LINDQUIST, D.A.,
     Application of the sterile-insect release technique in Mediterranean fruit fly suppression, J. Econ. Entomol.
     64 3 (1971) 708.
[2] RHODE, R.H., PERDOMO, A., DAXL, R., GUTIERREZ, J., Control of Mediterranean fruit flies in shade-
     grown coffee with ultra-low-volume aerial insecticide applications, J. Econ. Entomol. 65 6 (1972) 1749.

REPORT OF THE IIIrd MEETING OF THE INTERNATIONAL ORGANIZATION OF BIOLOGICAL CONTROL — WEST PALAEARCTIC REGION SECTION (IOBC/WPRS) WORKING GROUP ON GENETIC CONTROL OF THE MEDITERRANEAN FRUIT FLY

(IAEA, Vienna, 17 November 1973)

I.   Present: (A)   IOBC/WPRS Working Group

1.   Mellado (Spain, Co-ordinator)
2.   Cavalloro (C.E.C.)
3.   Cirio (Italy)
4.   Serghiou (Cyprus)
5.   Wakid (Egypt)

(B)   IOBC Members

1.   Simón (Peru)
2.   Turica (Argentina)

(C)   Observers

1.   Cavin (USDA, APHIS representing the USAID Pest Management Project)
2.   Chambers (USDA, ARS)
3.   Harris (USDA, ARS)
4.   Moore (FAO/IAEA, Rapporteur)

II.   Contents: Report

Annex I of Appendix II: List of laboratories involved in Mediterranean fruit fly activity.
Annex II of Appendix II: Addresses of participants in IOBC/WPRS medfly working group.

1.   The IOBC/WPRS Working Group on Genetic Control of the Medfly
during its 3rd meeting (IAEA, Vienna, 17 November 1973), recognized
that around the Mediterranean basin there were 11 laboratories involved
in research activities on medfly and/or mass production of the insect for
its suppression in their respective countries.
Mass production of medfly is also being carried out at the IAEA laboratory
at Seibersdorf.
The laboratories involved are listed in Annex I.

2.   The expertise accumulated in these various laboratories now has
reached the point where upscaling medfly production is no problem, given
the proper incentive and support. The Working Group, including the others
present at the meeting (addresses listed in Annex II), felt that any one of
these laboratories cannot on its own realistically undertake a large-scale
project of Mediterranean fruit fly control by the sterile-insect technique.
However, the experience of these laboratories shows that eradication of the
medfly by this technique under suitable conditions and with adequate support
is possible.

3.   Having surveyed conditions in the various Mediterranean areas where
the medfly is endemic, it was concluded that an eradication campaign should
be carried out through a concerted effort, provided this be concentrated
in a single appropriate area.   The co-ordination for such a collaborative
effort would have to come from an international organization and the group
agreed this could best be done by the Joint FAO/IAEA Division of Atomic
Energy in Food and Agriculture. The IOBC/WPRS Group, which originated
this collaborative concept, should play an active role in an advisory capacity
by providing:

    Advice on general guidelines of the campaign;
    Advice on implementation of the programme and selection of specialists;
    Expertise on location as required;
    Periodical meeting to review the progress of the programme and
    develop future activities.

4.   The review of the medfly situation in the Mediterranean region shows
that it is technically and economically feasible to achieve eradication of the
medfly in Cyprus. Geographically and environmentally, Cyprus has all
the conditions required to make such a project successful[1].

5.   The group understands that the Cyprus Government is interested in
such a project and would be ready to approach an organization such as the
World Bank, the European Communities, etc. for the financial support
needed.   The group would strongly endorse and support such a request
from the Cyprus Government.

---

[1] This report reflects the situation before 1974; the project has since been temporarily suspended.

6.   As regards implementation, the group is ready to contribute the expertise of all its members. Since this eradication campaign hinges on massive supplies of sterile medflies, the group suggests that the Cyprus Government makes a major effort to increase the present level of its medfly production by expanding the facility of its Agricultural Research Institute.

7.   The group recognizes that for eradication even with production at maximum capacity, the Cyprus facility could not meet the requirements in numbers of sterile medflies. Therefore, the group recommends that the necessary supplies of sterile medflies be obtained from the other laboratories in the Mediterranean area. Principal reliance should be on facilities logistically appropriate; the other facilities would provide emergency sources and genetic heterogeneity.

8.   The group also recommends that Cyprus creates the conditions needed for successful eradication such as quarantine, appropriate bait spraying and cultural practices, farmer participation and Government involvement as required.

9.   Success in this internationally co-ordinated campaign of eradication of the medfly in Cyprus would act as a catalyst for other programmes requiring international co-operation. Likewise, this would stimulate support of ongoing or proposed projects. Both these aspects would be major benefits of this proposal.

LABORATORIES INVOLVED IN MEDFLY ACTIVITIES

1. INIA, Madrid, Spain

2. Università degli Studi, Istituto di Entomologia Agraria, Palermo, Italy.

3. Laboratorio Applicazioni in Agricoltura, CNEN, Casaccia, 00060 Rome, Italy

4. Centro Regionale Anti-Insetti, Cagliari, Italy

5. Centro Studi Provinciale Antimalarico, Latina, Italy

6. Agricultural Research Institute, Nicosia, Cyprus

7. European Communities: Euratom, Common Nuclear Center, 21020 Ispra, Italy

8. Radiobiology Department, Atomic Energy Authority, Cairo, Egypt

9. Ministry of Agriculture, Giza, Egypt

10. Università degli Studi, Istituto di Entomologia Agraria, Portici, Napoli, Italy

11. Biological Control Institute, Citrus Marketing Board of Israel, Rehovot, Israel

12. IAEA Entomology Laboratory, IAEA, Vienna, Austria.

ADDRESSES OF PARTICIPANTS IN IOBC/WPRS MEDFLY WORKING GROUP

1. Dr. R. Cavalloro
    Commission of the European Communities,
    Ispra, Italy

2. Dr. G. Cavin
    USDA, Federal Center Blg.,
    Hyattsville, Maryland 20705, USA

3. Dr. D. L. Chambers
    Insect Attractants, Behavior and Basic Biology Research Laboratory,
    USDA, ARS,
    P.O. Box 14565,
    Gainesville, Florida 32601, USA

4. Dr. U. Cirio
    Centro di Studi Nucleari della Casaccia,
    Comitato Nazionale per l'Energia Nucleare,
    Casella Postale N. 2400,
    00100 Rome, Italy

5. Dr. E.J. Harris
    USDA, ARS, Western Region Southern California-Hawaii Area,
    Hawaiian Fruit Flies Lab.,
    P.O. Box 2280,
    Honolulu, Hawaii 96804, USA

6. Dr. L. Mellado
    Instituto Nacional de Investigaciones Agrarias,
    Avda. Puerta de Hierro,
    Madrid 3, Spain

7. Dr. I. Moore
    Joint FAO/IAEA Division of Atomic Energy in Food and Agriculture,
    Kärntner Ring 11,
    1011 Vienna, Austria

8. Dr. C. Serghiou
    Agricultural Research Institute,
    Ministry of Agriculture and Natural Resources,
    Nicosia, Cyprus

9. Dr. J.E. Simón F
    Centro Regional de Investigación Agraria,
    La Molina, Apartado 2791,
    Lima, Peru

10. Dr. A. Turica
    Instituto Nacional de Technología Agropecuaria,
    Centro de Investigaciones en Ciencias Agronómicas,
    C. C. 25, Castelar,
    Bueonos Aires, Argentina

11. Dr. A. M. Wakid
    Atomic Energy Establishment,
    Cairo, Arab Republic of Egypt

LIST OF PARTICIPANTS

BATEMAN, M.A.

C.S.I.R.O.,
School of Biological Sciences,
University of Sydney,
Sydney, N.S.W. 2006,
Australia

BOLLER, E.F.

Swiss Federal Research
    Station for Arboriculture,
    Viticulture and Horticulture,
Wädenswil, Switzerland

BRNETIC, D.

Institut za Jadranske Kulture
    i Melioraciju Krsa,
Split, Yugoslavia

BUSH, G.L.

Dept. of Zoology,
University of Texas,
Austin, TX 78712,
United States of America

CAVALLORO, R.

Commission of the European Communities,
Ispra, Italy

CAVIN, G.

USDA, Federal Center Blg.,
Hyattsville, MD 20705,
United States of America

CHAMBERS, D.L.

Insect Attractants,
Behavior and Basic Biology Research Lab.,
USDA, ARS,
P.O. Box 14565,
Gainesville, FL 32601
United States of America

CIRIO, U.

Laboratorio Applicazioni in Agricolture,
Comitato Nazionale per l'Energia Nucleare,
Casella Postale N. 2400,
00100 Rome, Italy

ECONOMOPOULOS, A.P.

Nuclear Research Center "Democritos",
Aghia Paraskevi,
Attiki, Greece

ENKERLIN S, D.   Instituto Tecnológico y de Estudios
                                        Superiores de Monterrey,
                                      Monterrey, N.L., Mexico

FISCHER-COLBRIE, P.   Bundesanstalt für Pflanzenschutz,
                                      Trunnerstrasse 5,
                                      1020 Vienna, Austria

GONZALEZ, R.   Plant Production and Protection Division,
                                      FAO,
                                      Rome, Italy

HAISCH, A.   Bayerische Landesanstalt für
                                        Bodenkultur und Pflanzenbau,
                                      Postfach,
                                      8 München 38,
                                      Federal Republic of Germany

HARRIS, E.J.   Hawaiian Fruit Flies Lab.,
                                      USDA, ARS,
                                      P.O. Box 2280,
                                      Honolulu, HI 96804,
                                      United States of America

MELLADO, L.   Instituto Nacional de
                                        Investigaciones Agrarias,
                                      Avda. de Puerta de Hierro,
                                      Madrid 3, Spain

REJESUS, R.S.   College of Agriculture,
                                      University of The Philippines at
                                        Los Baños,
                                      College, Laguna, Philippines

RUSS, K.   Bundesanstalt für Pflanzenschutz,
                                      Trunnerstrasse 5,
                                      1020 Vienna, Austria

SERGHIOU, C.   Agricultural Research Institute,
                                      Ministry of Agriculture and
                                        Natural Resources,
                                      Nicosia, Cyprus

SIMÓN F, J.E.   Dirección General de Investigaciones
                                        Agropecuarias,
                                      La Molina, Apartado 2791,
                                      Lima, Peru

THEUNISSEN, J.   Institute for Phytopathological Research,
                                      Binnenhaven 12,
                                      Wageningen, The Netherlands

TSITSIPIS, J.	Nuclear Research Center "Democritos", Aghia Paraskevi, Attiki, Greece
TURICA, A.	Instituto Nacional de Technología Agropecuaria, Centro de Investigaciones en Ciencias Agronómicas, C.C. 25, Castelar, Buenos Aires, Argentina
VALLO, V.	Institute of Experimental Phytopathology and Entomology, Ivanka pri Dunaji, Bratislava, Czechoslovakia
WAKID, A.M.	Radiobiology Dept. Atomic Energy Establishment, Cairo, Egypt
ZERVAS, G.	Nuclear Research Center "Democritos", Aghia Paraskevi, Attiki, Greece
ZIMMERMANN, E.	Bundesanstalt für Pflanzenschutz, Trunnerstrasse 5, 1020 Vienna, Austria
PROKOPY, R.	Nuclear Research Center "Democritos", Aghia Paraskevi, Attiki, Greece

International Atomic Energy Agency

LINDQUIST, D.A. SIGURBJÖRNSSON, B.	Joint FAO/IAEA Division of Atomic Energy in Food and Agriculture
BUTT, B.A. NADEL, D.J. TISCHLINGER, Anne WEHRSTEIN, A.	Division of Research and Laboratories
MOURAD, A.	IAEA Fellow at the Seibersdorf Laboratory

SCIENTIFIC SECRETARY

MOORE, I.	Joint FAO/IAEA Division of Atomic Energy in Food and Agriculture

HOW TO ORDER IAEA PUBLICATIONS

Exclusive sales agents for IAEA publications, to whom all orders and inquiries should be addressed, have been appointed in the following countries:

UNITED KINGDOM	Her Majesty's Stationery Office, P.O. Box 569, London SE 1 9NH
UNITED STATES OF AMERICA	UNIPUB, Inc., P.O. Box 433, Murray Hill Station, New York, N.Y. 10016

In the following countries IAEA publications may be purchased from the sales agents or booksellers listed or through your major local booksellers. Payment can be made in local currency or with UNESCO coupons.

ARGENTINA	Comisión Nacional de Energía Atómica, Avenida del Libertador 8250, Buenos Aires
AUSTRALIA	Hunter Publications, 58 A Gipps Street, Collingwood, Victoria 3066
BELGIUM	Service du Courrier de l'UNESCO, 112, Rue du Trône, B-1050 Brussels
CANADA	Information Canada, 171 Slater Street, Ottawa, Ont. K 1 A OS 9
C.S.S.R.	S.N.T.L., Spálená 51, CS-110 00 Prague
	Alfa, Publishers, Hurbanovo námestie 6, CS-800 00 Bratislava
FRANCE	Office International de Documentation et Librairie, 48, rue Gay-Lussac, F-75005 Paris
HUNGARY	Kultura, Hungarian Trading Company for Books and Newspapers, P.O. Box 149, H-1011 Budapest 62
INDIA	Oxford Book and Stationery Comp., 17, Park Street, Calcutta 16
ISRAEL	Heiliger and Co., 3, Nathan Strauss Str., Jerusalem
ITALY	Libreria Scientifica, Dott. de Biasio Lucio "aeiou", Via Meravigli 16, I-20123 Milan
JAPAN	Maruzen Company, Ltd., P.O.Box 5050, 100-31 Tokyo International
NETHERLANDS	Marinus Nijhoff N.V., Lange Voorhout 9-11, P.O. Box 269, The Hague
PAKISTAN	Mirza Book Agency, 65, The Mall, P.O.Box 729, Lahore-3
POLAND	Ars Polona, Centrala Handlu Zagranicznego, Krakowskie Przedmiescie 7, PL-00-068 Warsaw
ROMANIA	Cartimex, 3-5 13 Decembrie Street, P.O.Box 134-135, Bucarest
SOUTH AFRICA	Van Schaik's Bookstore, P.O.Box 724, Pretoria
	Universitas Books (Pty) Ltd., P.O.Box 1557, Pretoria
SPAIN	Nautrônica, S.A., Pérez Ayuso 16, Madrid-2
SWEDEN	C.E. Fritzes Kungl. Hovbokhandel, Fredsgatan 2, S-103 07 Stockholm
U.S.S.R.	Mezhdunarodnaya Kniga, Smolenskaya-Sennaya 32-34, Moscow G-200
YUGOSLAVIA	Jugoslovenska Knjiga, Terazije 27, YU-11000 Belgrade

Orders from countries where sales agents have not yet been appointed and requests for information should be addressed directly to:

Publishing Section,
International Atomic Energy Agency,
Kärntner Ring 11, P.O.Box 590, A-1011 Vienna, Austria

75- 08192